Urban Agriculture

Series editors

Christine Aubry, AgroParisTech, INRA UMR SADAPT, Paris, France
Éric Duchemin, Université du Québec à Montréal, Institut des Science de
l'Environnement, Montreal, Québec, Canada
Joe Nasr, Centre for Studies in Food Security, Ryerson University,
Toronto, Ontario, Canada

W0235047

The *Urban Agriculture* Book Series at Springer is for researchers, professionals, policy-makers and practitioners working on agriculture in and near urban areas. Urban agriculture (UA) can serve as a multifunctional resource for resilient food systems and socio-culturally, economically and ecologically sustainable cities.

For the Book Series Editors, the main objective of this series is to mobilize and enhance capacities to share UA experiences and research results, compare methodologies and tools, identify technological obstacles, and adapt solutions. By diffusing this knowledge, the aim is to contribute to building the capacity of policy-makers, professionals and practitioners in governments, international agencies, civil society, the private sector as well as academia, to effectively incorporate UA in their field of interests. It is also to constitute a global research community to debate the lessons from UA initiatives, to compare approaches, and to supply tools for aiding in the conception and evaluation of various strategies of UA development.

The concerned scientific field of this series is large because UA combines agricultural issues with those related to city management and development. Thus this interdisciplinary Book Series brings together environmental sciences, agronomy, urban and regional planning, architecture, landscape design, economics, social sciences, soil sciences, public health and nutrition, recognizing UA's contribution to meeting society's basic needs, feeding people, structuring the cities while shaping their development. All these scientific fields are of interest for this Book Series. Books in this Series will analyze UA research and actions; program implementation, urban policies, technological innovations, social and economic development, management of resources (soil/land, water, wastes…) for or by urban agriculture, are all pertinent here.

This Book Series includes a mix of edited, coauthored, and single-authored books. These books could be based on research programs, conference papers, or other collective efforts, as well as completed theses or entirely new manuscripts.

More information about this series at http://www.springer.com/series/11815

Caroline Brand • Nicolas Bricas
Damien Conaré • Benoit Daviron
Julie Debru • Laura Michel
Christophe-Toussaint Soulard
Editors

Designing Urban Food Policies

Concepts and Approaches

 Springer

OPEN

Editors

Caroline Brand
Chaire Unesco Alimentations du monde
Montpellier SupAgro
MONTPELLIER CEDEX 5, France

Nicolas Bricas
UMR MOÏSA
Cirad
MONTPELLIER CEDEX 5, France

Damien Conaré
Chaire Unesco Alimentations du monde
Montpellier SupAgro
MONTPELLIER CEDEX 5, France

Benoit Daviron
UMR MOÏSA
Cirad
MONTPELLIER CEDEX 5, France

Julie Debru
Chaire Unesco Alimentations du monde
Montpellier SupAgro
MONTPELLIER CEDEX 5, France

Laura Michel
Faculty of Law and Political Science
University of Montpellier
MONTPELLIER CEDEX 2, France

Christophe-Toussaint Soulard
UMR Innovation
INRA
MONTPELLIER CEDEX 2, France

First published in 2017 by Éditions Quæ; Construire des politiques alimentaires urbaines: Concepts et démarches

ISSN 2197-1730 ISSN 2197-1749 (electronic)
Urban Agriculture
ISBN 978-3-030-13960-5 ISBN 978-3-030-13958-2 (eBook)
https://doi.org/10.1007/978-3-030-13958-2

This Springer imprint is published by the registered company Springer Nature Switzerland AG.
The registered company address is: Gewerbestrasse 11, 6330 Cham, Switzerland

Foreword

The urban food system sustainability issue first and foremost raises personal concerns regardless of whether you live in a city or rural area, in a so-called poor or rich country, or are a leader or an ordinary citizen. One may ask: Under what conditions can I feed my family? What part depends on my culture, my economic capacity, my personal choices or the prevailing food system? How would I adapt to a crisis situation?

These personal questions soon shift the spotlight onto the systemic dimension of urban food systems and their sustainability. One may then ask: Have I ever lived in a nonsustainable urban food system? Have I been aware of such systems? Individual responses to these questions would be wide-ranging, but several images come to mind: the food plight of people living in cities during wartime, the impact of the 2014–2015 Ebola epidemic in West Africa on the distribution of foodstuffs and the urban riots that gave rise to the Arab Spring in 2010, which were triggered by uncontrolled soaring food prices. Nowadays in Europe, commodity producers are also bearing the brunt of the economic power of supermarket chains—and, who knows, next it could be consumers in that position. In the summer of 1982, meat supplies to a large European capital were at a standstill—but city dwellers managed to mobilize their family networks in rural areas within 2 weeks, leading to the formation of informal supply chains to solve the problem.

There is a risk of feeling powerless when defining an urban food system management strategy because of the complexity of the issue and the diversity of situations, constraints, stakeholders and viewpoints. Experience sharing, awareness of analytical approaches and mobilizable reference frameworks are thus particularly crucial in this setting.

Some urban systems—including transportation and water provisioning—are amenable to technical analysis or systems modelling, while scenario simulation can a priori help to identify weaknesses and preventively implement improvement strategies.

Urban food systems are however much more decentralized, so the authors of this book advocate a sustainability approach geared towards strengthening actors' expertise, their ability to explain their vision and objectives, identify levers to which they have access and implement them in co-governance situations. This type of approach is favoured by the Milan Urban Food Policy Pact that was signed in Milan in October 2015 by more than 100 mayors of cities from around the world.

This approach is jointly driven by historical knowledge, which can clarify current situations and issues, and by an in situ case study around the city of Montpellier (France). The main recent theoretical approaches presented in Chap. 4 and the conceptual framework proposed in Chap. 5 will help many urban food system stakeholders analyse the setting of their city, identify similar experiences, as well as share, compare and capitalize on knowledge.

Given the major challenge of ensuring the sustainability of urban food systems, this book—coordinated by the UNESCO Chair on World Food Systems, CIRAD and INRA, with the support of Agropolis Fondation—provides a conceptual framework and basis for knowledge development and sharing. It comes as a very welcome contribution on this crucial issue.

Director, Agropolis Fondation   Pascal Kosuth
Montpellier, France

Preface

Cities have been propelled into the forefront of the debate about sustainable food policies because, with rapid urbanization, the urban realm has become a locus for three of the most significant challenges to the conventional food system—multifunctionality, co-governance and city-regionalism.

Under the banner of multifunctionality, the urban food movement has challenged the conventional idea that food should be treated as a simple commodity like any other. But urban food campaigners resist this idea by arguing that food cannot be so easily commodified because, unlike manufactured commodities, we ingest food and it plays a vital role in the health and wellbeing of people and the planet.

The urban food movement—a loose and sometimes chaotic assemblage of municipal activism and civic engagement—is also challenging the idea that the food policy arena is reserved for corporate interests, national governments and international bodies like the World Trade Organization (WTO). One of the most rapidly growing social movements of our time is the advent of food policy councils, or food policy partnerships, where civil society organizations are joining forces with municipal politicians and officers to fashion a more sustainable urban foodscape, one in which the values of public health, social justice and ecological integrity are treated more seriously in the food policy equation. New forms of co-governance are beginning to emerge in our cities as politicians realize that they have to design policies *with* rather than for civil society—and urban food policy is in the forefront of this process.

City-region food systems constitute a third challenge to the conventional food policy mindset, which has been dominated until recently by the logic of globalization and the placeless foodscapes where price is extolled over provenance. The new urban food movement champions a sustainable food system in which cities are able to reconnect with their regional hinterlands as well as consuming fairly traded products from afar. Sustainability should not be confused with green autarchy! Sustainable city-region food strategies pose a challenge because they contest one of the deepest divisions in capitalist society—between town and country.

Creating sustainable urban food policies will face many barriers, not least because they pose such fundamental challenges to vested interests in the conventional food system. But this system is not set in aspic, and one of the great merits of this book—based on the work of leading scholars associated with the Agropolis Foundation—is that it explores the scope for sustainable urban food policies, one of the greatest societal challenges of the twenty-first century.

Professor of Governance and Development,   Kevin Morgan
School of Geography and Planning
Cardiff University
Cardiff, Wales, UK

Acknowledgements

We would like to thank all those who have participated in the Sustainable Urban Food Systems (SurFood) programme, thus contributing to the emergence of a collective multidisciplinary vision of sustainable urban food and of projects supported by the Montpellier-based scientific community in collaboration with local stakeholders. This programme brought together over 100 researchers, lecturers and engineers of the research units: ART-DEV, CEPEL, CIRED, LET, MRM, NORT, NutriPass, Recycling and Risk and TETIS, in France; ENDA-Graf, IPAR, ISRA and UCAD in Dakar (Senegal); HUST and MALICA (CASRAD and FAVRI) in Hanoi (Vietnam); ETAN in Kenitra (Morocco); IAV and INAU in Rabat (Morocco); as well as Bioversity International. We are grateful to Florence Egal (Consultant, Milan Urban Food Policy Pact) and Anna Faucher (International Urban Food Network) for their contributions to the brainstorming workshops. We also thank Françoise Jarrige, Ronan Le Velly and Simon Vonthron for their work regarding metropolitan Montpellier, along with Valérie de Saint Vaulry, Anne-Sophie Muepu, Sam Sandiani and Isabelle Touzard of Montpellier Méditerranée Métropole.

This book could not have been published without the support of Agropolis Fondation which initiated and backed the SurFood programme, as well as the Daniel and Nina Carasso Foundation which has been supporting the UNESCO Chair on World Food Systems since 2012.

Finally, we thank David Manley and Paul Cowan for translating this book.

Introduction

The present book is in keeping with the aims of the Sustainable Urban Food Systems (SurFood) research project supported by Agropolis Fondation. This flagship project was set up to coordinate, amplify and globally showcase the interdisciplinary research of the different institutional members of Labex Agro. It has been written in response to a request from Agropolis Fondation's Science Council to build a conceptual framework for the analysis, evaluation and development of sustainable urban food systems.

The Twenty-First Century Heralds the Dawn of a New Era for Cities

Following the reign of nation states, there is every reason to believe that the twenty-first century will be a new era for cities.

First because, from a structural standpoint, currently over half of the world population is urban (compared to 30% in 1950), and it will increase to two-thirds by 2050 according to the United Nations (2014). Medium-sized cities have considerable weight in this process since half of the urban population presently lives in cities of fewer than 500,000 inhabitants, while 1 urban dweller in 8 lives in a megacity of more than 10 million inhabitants. Africa and Asia—two continents that are still mainly rural—will be the locus of most of this urban population growth (2.5 billion more people in the next 35 years). Together, China, India and Nigeria will account for nearly 40% of the world's population growth by 2050. This raises major challenges in meeting housing, infrastructure, transportation, energy, employment, education, health and, of course, food needs.

Second because cities are gaining tremendous social, political and economic power. This power rises which—in addition to the demographic weight that cities represent—may be partly explained by production system changes taking place worldwide in a globalization setting and by the financial disengagement of States in land-use planning. Cities represent powerful local hubs that States can rely on to manage transitions to new development models. Cities have thus extended and asserted their power in many areas of social life to transform an ambient 'ecodesire' into tangible local reality (Haëntjens 2009) while developing their scope of operations to ensure their sustainability (Emelianoff 2007).

Cities as 'Human Settlements'

This trend is reflected by the growing number of territorial sustainable development policies and the development of global networks of local urban governments. At the 1992 United Nations Conference on Environment and Development (Rio Earth Summit), local authorities in each country were called upon to set up local Agenda 21 programmes tailored to their local setting, considering that: '[…] As the level of governance closest to the people, they play a vital role in educating, mobilizing and responding to the public to promote sustainable development' (Chapter 28 on local authorities' initiatives in support of Agenda 21).

Urban networks (Metropolis, International Council for Local Environmental Initiative and the Global Network of Cities, Local and Regional Governments) have spearheaded initiatives to ensure the tangible implementation of local Agenda 21 programmes.

The sustainability objectives were strengthened 4 years later at the Second United Nations Conference on Human Settlements (Habitat II) in Istanbul in 1996. Launched 20 years earlier in Vancouver in 1976, this UN process marked the beginning of an initiative to provide a reference framework for international cooperation on the topic of 'human settlements'. This expression reflects a desire to consider cities as the outcome of relationships between inhabitants and their structures—a form that urban populations build and organize (Emelianoff 2007).

Many voluntary partnerships between cities worldwide were later formed, especially on the climate change issue, for the exchange of good practices and joint commitments. These networks have often played a major underlying yet little-known role in international governance. Given, for instance, that cities account for around 70% of total greenhouse gas (GHG) emissions even though they only occupy 2% of the Earth's surface area (UN-Habitat, 2011), the C40 Cities Climate Leadership Group (a network of 83 cities worldwide) has made commitments to reduce emissions by 2020, corresponding to the cumulated emissions of Portugal and Argentina. Organized in networks, cities are thus now managing to cope with a number of global challenges.

Global and Organic Cities

Cities have also gained economic weight in this new globalization phase. A World Bank analysis (Kilroy et al. 2015) showed that three-quarters of 750 surveyed cities had grown faster than their respective national economies since the 2000s.

The American geographer, Saskia Sassen, devised the 'global city' concept to describe the impact of liberalization and financialization processes on some cities since the 1980s (Sassen 1991). According to her, the current globalization phase is marked by a simultaneous geographic dispersal and concentration dynamic, whereby economic activities and industrial installations are increasingly dispersed throughout the world, while central management functions (accounting, legal, public relations, etc.) have become highly reconcentrated in a number of global cities that host the headquarters of major multinational companies, financial institutions and stock markets. These global cities are therefore defined by the key role they play—with central command functions—in the new global economy, which has given rise to a transnational urban system. The spatial scale of these global cities, which are benefitting from privatization and economic deregulation, is thus taking ascendance over the national scale.

Globalization has cut main cities off from their national economies, or even distanced them from their territorial anchorage. The food sector therefore represents a tremendous opportunity for cities to recreate the sometimes lost link with the environments that produce their food. It is also a chance to forge new links with 'organic cities' of the past (Steel 2008), which prevailed prior to the industrial revolution when cities were literally shaped by food—with the spatial structure of cities organized around central food markets, often surrounded by buildings symbolizing the religious (church), economic (chamber of commerce) and political (city hall) powers, with street names referring to foods or food shops and the architectural design of houses related to food supply, etc.

Feeding cities has therefore become crucial despite the fact that this issue was no longer being taken seriously in some regions of the world. Carolyn Steel takes the example of London, where it seems to be taken for granted that sufficient food must be produced, transported, distributed, sold and cooked daily to feed 8.6 million people, and the generated waste must also then be managed. This challenge nevertheless must be met every day in all cities worldwide.

Feeding Cities: An Urban Planning Issue

There is rising awareness that growing and often poorly controlled urbanization leads to urban sprawl, socio-spatial inequality, pollution and environmental degradation associated with nonsustainable modes of production and consumption. The increased distancing—geographic (remoteness from basins), economic (increased number of intermediaries) and cognitive (ignorance of production conditions)—between cities and supply basins raises many problems: increased transport costs, energy consumption and food loss and wastage. Finally, relationships between city and rural dwellers are becoming less tight-knit as a result of the many food processing, logistics, distribution and catering operations.

Food has again become a global discussion issue as a result of the 2008 crisis regarding agricultural raw material prices and following numerous health crises (bovine spongiform encephalopathy, avian influenza, etc.), while cities are increasingly interested in finding ways to meet city dwellers' expectations on improving their diet. This twofold global/local movement is reflected on a territorial level by an increase in initiatives on food relocalization, urban agriculture, farmland protection, school canteen provisioning, etc. This plethora of innovations is still poorly structured, while accounting for or including it in integrated food policies is still a recent phenomenon.

Pothukuchi and Kaufman (1999) were among the first authors to focus on the importance of the role of food in the city. According to these authors, at the time, there were at least four reasons for city representatives' lack of interest in the food issue: the food system did not require special attention as it was considered to be functioning well, the food sector was not within the purview of urban planners, this sector (contrary to the transport and housing sectors) did not attract financing, and, finally, food was considered to be primarily a rural agricultural issue (not an urban one).

According to Morgan (2009), the latter argument is not admissible to justify the 'puzzling omission' on the part of planners regarding food. First, the multidimensional aspect of the food system means that it has a substantial impact on other sectors such as public health, social justice, energy, water, land, transport and economic development. All of these sectors are considered to be key concerns of urban representatives who have every right to deal with them. Second, considering food production as an exclusively rural activity challenges the fact that in many cities worldwide, urban agriculture has a pivotal role in food security and in others it inspires a rich socioeconomic movement geared towards producing food in cities.

Emergence of New Urban Food Strategies

The food issue—certainly 'too big to see' (Steel 2008)—has long been overlooked by urban actors. Over the last two decades, however, many cities have developed their own food strategy while incorporating different aspects of the system in a common framework: production, processing, distribution, access, consumption and waste management. These strategies are often integrated in broader initiatives to promote urban sustainability (Jennings et al. 2015) and reduce the urban-rural divide (Forster and Getz Escudero 2014).

To this end, local urban governments (cities, conurbations, metropolises) have a number of levers at hand: tenders for collective catering food supplies (school canteens, hospitals and other public institutions), land management (especially preservation of production areas), development of commercial areas, neighbour-hoods and flows, formation of governance structures such as food policy councils, creation of farmers' markets, etc. A local food strategy along these lines could help in dealing with different issues: improving food quality, reducing losses and wast-age, stimulating local economic development, strengthening solidarity between urban dwellers, improving their health, protecting the environment, etc. (Wiskerke 2009).

Strategies implemented by local urban governments are supported by a broader movement of initiatives geared towards relocalizing food systems run by individual or associative bodies, local governments at different scales, farmers, agrifood com-panies and research stakeholders (Feanstra 1997, 2002; Hendrickson and Heffernan 2002; Feagan 2007; Muchnik et al. 2007; Martinez et al. 2010; *Association des Régions de France* 2014; Rastoin 2016). This food relocalization movement has also been the brunt of criticism, e.g. accused of being a 'local trap' (Born and Purcell 2006)—the claim that the geographical governance scale alone does not ensure food system sustainability. Indeed, the actors and the agenda they set up are the actual guarantors (Hinrichs et al. 1998; Brown and Purcell 2005; Watts et al. 2005; Born and Purcell 2006; Kneafsey 2010).

Urban areas in North America have been testing food policies since the early 1990s (Neuner et al. 2011; Viljoen and Wiskerke 2012), with Toronto being one of the cities spearheading this trend (Blay-Palmer 2009), but this dynamic is not solely embraced by industrialized countries. The Brazilian metropolis Belo Horizonte is thus lauded as a pioneer for having developed an urban policy to combat food inse-curity. Its experience provided a model for building the Brazilian nationwide *Fome Zero* (zero hunger) programme (Rocha and Lessa 2009). In the same vein, many other cities have experimented with local food insecurity control strategies via school canteens, urban agriculture and short supply chains, including (among many other examples) Mexico City, Medellin (Colombia), Rosario (Argentina), Gampaha (Sri Lanka), Nairobi (Kenya) and Accra (Ghana).

Civil society mobilization has been decisive in all cases (Morgan 2009). These initiatives provide technical and economic alternatives to the standard food produc-tion, processing and distribution model while also giving rise to other modes of governance that stimulate local democracy processes (Guthman 2008; Lang et al. 2009; Starr 2010; Block et al. 2012).

These local initiatives are gaining momentum and converging in national and international networks such as Sustainable Food Cities (UK), Resource Centres on Urban Agriculture and Food Security (RUAF Foundation), Transition Network (TN), African Food Security Urban Network (AFSUN), Eating City, International Council for Local Environmental Initiatives (ICLEI), International Urban Food Network (IUFN), Sustainable Food Planning Group of the Association of European Schools of Planning (AESOP), etc.

Caroline Brand (2015) nevertheless clearly demonstrated that the movement of territories towards an integrated global view on the food issue is not a clear-cut trend. There is still often a gap between the rhetoric on the issue and the actual fact of taking the food situation into account in territories. Moreover, the strategies implemented are often fragmentary and just deal with one of the many facets of the food system, or else they do not seek to explicitly intervene on the food system per se (Hodgson 2012).

Towards a Conceptual Framework

This book is structured as follows for the purpose of gaining insight into, analysing and conceptualizing the dynamics between cities and their food systems:

It first focuses on the limitations of the industrialized food system (Chap. 1), which is currently dominant worldwide but nonsustainable. Many reports have highlighted the pitfalls: high nonrenewable resource consumption, socially and economically unfair, not very rewarding for some stakeholders, a major contributor to biodiversity loss and climate change via GHG emissions and a generator of food waste and wastage. Global urbanization magnifies these negative impacts. Meanwhile, however, cities pool a number of resources that provide an opportunity to enhance the sustainability of urban food systems.

The relationship between cities and food is then discussed from a historic standpoint, from antiquity to World War II, while being geographically focused on Europe and the Mediterranean region (Chap. 2). Different types of intervention by urban authorities are analysed over a time scale: safeguarding of food supply chains, organization of local markets, ensuring food quality, combating food fraud, waste and soil fertility management, etc. The aim of this chapter is therefore to showcase the wealth of initiatives undertaken over time to promote urban food policies so as to broaden the scope of discussion options in current debates.

Nowadays, to tackle the issues set out in Chap. 1, what policies have been implemented by cities, and what are the governance aims, means and models (Chap. 3)? This chapter—via a typology of urban food polices on these points—investigates the relevance of urban food supply relocalization and seeks to measure the extent to which food systems could be effectively modified at the local scale.

The sustainability of food systems and especially of urban food policies has become a focus of research along with the development of a number of conceptual frameworks for their analysis (Chap. 4). This chapter provides an overview of existing frameworks. While not striving to be exhaustive, a few emblematic approaches that have been developed are presented to take the urban food issue into account in an integrated way. This chapter highlights the need to combine the different approaches.

This appraisal provides an opportunity to present the approach developed by the SurFood programme, which aims to develop urban food policies and build analysis and assessment tools, while providing knowledge on possible policy options that urban actors could potentially implement to build more sustainable urban food system models (Chap. 5). This approach dovetails three dimensions: levers that cities have at hand to manage food at their scale, the problems they must deal with and the governance arrangements implemented to launch a food strategy. The objectives of this approach, its uses and potential applications are specified in the chapter.

This approach has been applied in the metropolitan Montpellier area. In 2014, Montpellier Méditerranée Métropole voted in an agroecology and food policy with the support of researchers involved in the SurFood project (Chap. 6). This chapter investigates the urban food governance emergence process, from the point where food is placed on the agenda when drawing up public policies in local urban governments to the mobilization of civil society actors.

Finally, the book concludes by presenting a series of potential avenues of research to gain insight into and support the transition dynamics needed to achieve more sustainable urban food systems.

UNESCO Chair in World Food System,   Damien Conaré
Montpellier SupAgro, Montpellier, France

References

Association des Régions de France (2014) Déclaration de Rennes : pour des systèmes alimentaires territorialisés. http://regions-france.org/wp-content/uploads/2016/10/déclaration-finale.pdf. Accessed 25 Nov 2016

Blay-Palmer A (2009) The Canadian pioneer: the genesis of urban policy in Toronto. Int Plan Stud 14(4):401–416

Block DR, Chávez N, Allen E, Ramirez D (2012) Food sovereignty, urban food access, and food activism: contemplating the connections through examples from Chicago. Agric Hum Values 29:203–215

Born B, Purcell M (2006) Avoiding the local trap: scale and food systems in planning research. J Plan Edu Res 26(2):195–207

Brand C (2015) Alimentation et métropolisation : repenser le territoire à l'aune d'une problématique vitale oubliée. Doctoral thesis, Geography specialization, Université Grenoble Alpes, Laboratoire PACTE, UMR 5194, École doctorale 454 Science de l'homme, du politique et du territoire, Grenoble, 659 p

Brown JC, Purcell M (2005) There's nothing inherent about scale: political ecology, the local trap, and the politics of development in the Brazilian Amazon. Geoforum 36:607–624

Emelianoff C (2007) La ville durable : l'hypothèse d'un tournant urbanistique en Europe. L'information géographique 71(3):48–65

Feagan R (2007) The place of food: mapping out the 'local' in local food systems. Progr Hum Geogr 31(1):23–42

Feenstra G (1997) Local food systems and sustainable development. Am J Altern Agric 12(1):28–37

Feenstra G (2002) Creating space for sustainable food systems: lessons from the field. Agric Hum Values 19(2):99–106

Forster T, Getz Escudero A (2014) City Regions as Landscapes for People, Food and Nature. EcoAgriculture Partners, on behalf of the Landscapes for People, Food and Nature Initiative, Washington, DC

Guthman J (2008) Bringing good food to others: investigating the subjects of alternative food practice. Cult Geogr 15:431–447

Haëntjens J (2009) Le pouvoir des villes ou l'art de rendre désirable le développement durable. Collection Monde en cours, Nouvelles éditions de l'Aube, 158 p

Hendrickson MK, Heffernan WD (2002) Opening spaces through relocalization: locating potential resistance in the weaknesses of the Global Food System. Sociol Rural 42(4):347–369

Hinrichs C, Kloppenburg J, Stevenson S, Lezberg S, Hendrickson J, DeMaster K (1998) Moving beyond global and local. United States Department of Agriculture, Regional Research Project NE-185 working statement, October 2

Hodgson K (2012) Planning for food access and community-based food systems: a national scan and evaluation of local comprehensive and sustainability plans. American Planning Association

Jennings S, Cotte J, Curtis T, Miller S (2015) Food in an Urbanised World – The Role of City Region Food Systems in Resilience and Sustainable Development. 3Keel

Kilroy A, Francis L, Mukim M, Negri S (2015) Competitive cities for jobs and growth: what, who, and how. World Bank Group, Washington DC

Kneafsey M (2010) The region in food – important or irrelevant? Camb J Reg Econ Soc 3(2):177–190

Lang T, Barling D, Caraher M (2009) Food policy: integrating health, environment and society. Oxford University Press, Oxford

Martinez S, Hand M, Da Pra M, Pollack S, Ralston K, Smith T, Vogel S, Clark S, Lohr L, Low S, Newman C (2010) Local food systems: concepts, impacts, and issues. ERR 97, U.S. Department of Agriculture, Economic Research Service, May

Morgan K (2009) Feeding the city: the challenge of urban food planning. Int Plan Stud 14(4):341–348

Muchnik J, Requier-Desjardins D, Sautier D, Touzard JM (2007) Introduction aux SYAL. Économies et Sociétés 29:1465–1485

Neuner K, Kelly S, Samina R (2011) Planning to eat? Innovative local government plans and policies to build healthy food systems in the United States. Healthy Kids-Healthy Communities-Buffalo partnership and the Food Systems Planning and Healthy Communities Lab, University at Buffalo

Pothukuchi K, Kaufman JL (1999) Placing the food system on the urban agenda: The role of municipal institutions in food systems planning. Agric Hum Values 16(2):213–224

Rastoin J-L (2016) Les systèmes alimentaires territorialisés: enjeux et stratégie de développement. J Resolis 7:12–18

Rocha C, Lessa I (2009) Urban governance for food security: the alternative food system in Belo Horizonte, Brazil. Int Plan Stud 14(4):389–400

Sassen S (1991) The global city: New York, London, Tokyo. Princeton University Press, Princeton. 480 p

Starr A (2010) Local food: a social movement? Cult Stud Crit Methodol 10(6):479–490

Steel C (2008) Hungry city: how food shapes our lives. Random House Group Ltd., London. 400 p

UN-Habitat (2011) Cities and Climate Change: Global Report on Human Settlements 2011. Earthscan, London/Washington DC. 300 p

Viljoen A, Wiskerke JS (2012) Sustainable food planning: evolving theory and practice. Wageningen Academic Publishers, Wageningen

Watts DCH, Ilbery B, Maye D (2005) Making reconnections in agro-food geography: alternative systems of food provision. Prog Hum Geogr 29(1):22–40

Wiskerke JSC (2009) On places lost and places regained: reflections on the alternative food geography and sustainable regional development. Int Plan Stud 14(4):369–387

Contents

Urbanization Issues Affecting Food System Sustainability

Nicolas Bricas with Sabine Barles, Gilles Billen, and Jean-Louis Routhier
(Box Contributors)

This chapter identifies urbanization issues affecting food system sustainability. Since the nineteenth century and the onset of industrialization, cities have been both the product and motor of food systems, which are expanding worldwide despite the impression that they are nonsustainable. Since both the problems and resources are concentrated in cities, the latter are also a source of innovation, which can in turn help enhance the sustainability of food systems.

Cities, Sustainability and Food Systems: Definitions

Cities

To understand sustainability issues that cities may cause or could help solve, cities may be viewed through several features:

- First the city is a concentration of human beings, a space or habitat with a highly dense population, thus limiting areas for extensive agricultural production. Cities are essentially nurtured by external areas. This population concentration often gives rise to traffic problems given the density of goods needed by city dwellers. Originally a migration focal point, the city is also a space of coexistence and interaction between communities of different geographical and cultural origins (Lamizet 2007)—a source of both innovations and possible tensions.
- The city is also a concentration of resources—power, knowledge, wealth and biomass—contrasting with the scattered resources in rural areas. This generates power asymmetry between rural and urban environments. Some authors consider

N. Bricas (✉)
UMR MOÏSA, Cirad, MONTPELLIER CEDEX 5, France
e-mail: nicolas.bricas@cirad.fr

© The Author(s) 2019
C. Brand et al. (eds.), *Designing Urban Food Policies*, Urban Agriculture,
https://doi.org/10.1007/978-3-030-13958-2_1

1

that cities develop through perpetual exploitation of rural areas (Salomon Cavin and Marchand 2010; Salomon Cavin and Mathieu 2014), while others point out that cities have been the drivers of agricultural development, offering outlets for higher value-added products (Tacoli 1998; Cour 2004).

– The city is a space of commercial and cultural exchange, with the urban market being a focal point of trade flows, particularly of agricultural goods. It hence represents a network hub whose ramifications can extend way beyond the outskirts of the urban area.

In addition to these three features that have been always been inherent to urban centres, we should mention two other traits that have been specific to cities since the nineteenth century:

– Urban growth has been a product of industrialization for nearly two centuries. This is understood here as involving the development of a mode of production using nonrenewable resources, particularly for energy, rather than the development of manufacturing companies (Krausmann and Haberl 2002). But this industrialization process is also concentrated in cities. Urbanization and industrialization thus seem to be closely linked, especially since issues related to these two phenomena are inseparable.

– The concentration of wealth and power has accelerated as a result of the major productivity gains achieved through industrialization. Industrial cities are spaces marked by high inequality between rich and poor populations living in close proximity, between small and large companies operating in the same markets, and between central governments where industrial, financial and political interests often overlap. The growing inequality is reflected in the marginalization of part of the population due to exclusion from the production system where mechanization is increasingly replacing human labour, thus constituting a reservoir of cheap manpower.

The evolution in the relationship the city has with its food system can be interpreted as a case of multifaceted distancing (Bricas et al. 2013):

– geographical distancing with remote relocation of supply areas because of urban expansion and facilitated by the decrease in fossil fuel based transportation costs;

– economic distancing with the multiplication of intermediaries between agricultural producers and consumers for food dissemination, processing, storage and distribution;

– cognitive distancing with a growing share of the population born in the city as it ages—a population that is unfamiliar with the rural community, with the virtual disappearance of contacts between city dwellers and farmers, and with knowledge of the agriculture and food spheres no longer being direct but instead mediated by science and information media;

– political distancing with citizens losing control of their food system—with their range of preferences being confined to choices between supply areas and products. Despite the development of consumerism, eaters feel deprived of power to orient the system, which is in the hands of a just few dominant stakeholders.

Food System

The food system represents a set of interrelated activities regarding the production, dissemination, processing and use of food, waste and the required resources. Food issues have long been overshadowed by production sufficiency concerns—a major focus of interest—but the food system concept is broader in scope, encompassing production and all that precedes and succeeds it (Malassis 1983; Rastoin and Ghersi 2010). This concept helps tackle issues specifically related to how agricultural products are used to feed populations (Sobal et al. 1998).

Still recently considered as resulting in consumption (from the Latin term *consumere)*, i.e. destruction, the food system should now incorporate the fate of the waste produced. This waste has historically been used as fertilizer in agricultural production cycles until being replaced by petroleum-based nitrogen fertilizers and mined phosphate and potassium, with the result that waste is no longer a food system concern (Lewin 2009). We include waste as a food system component with the aim of minimizing waste quantities and recycling the intractable part, thus promoting sustainability.

Food

Food is considered here from a multifunctional standpoint (Fischler 1990; Poulain 2002). It is not solely aimed at meeting human needs (nutritional function), it also serves to create and maintain social interactions through the organization of the movement of products (market role) and the organization of meals (commensality role). It also provides pleasure and thus has an artistic aspect, as promoted by gastronomy (hedonic function). Food has an important identity function since it is incorporated in the body and thus helps build individual and collective identities. The biological function seems vital, but can no longer be considered a priority or fundamental once biological survival is no longer threatened. The relative importance of food functions is generally culture-dependent.

Sustainability

There is current no consensus on what pertains to sustainability in contemporary society issues. The sustainability concept was developed to cope with irreversible environmental situations resulting from changes in modes of production, especially via industrialization. Then a social dimension soon emerged, including problems of inequality, in addition to health concerns and even cultural aspects. Many problems are nowadays dealt with in the name of sustainability (Godard 1994).

On the basis of definitions proposed by FAO and Bioversity International (Burlingame and Dernini 2010), as well as IPES Food (2015), here we have adopted a definition that considers food systems are sustainable when they:

- protect the environment and biodiversity without depleting nonrenewable resources and biodiversity, and without polluting;
- provide universal access to sufficient, healthy, nutritional and culturally acceptable food;
- rely on an inclusive economic system that creates jobs for everyone while reducing power inequalities between businesses and within value chains so as to ensure a more equitable distribution of added value;
- encourage social cohesion and respect cultural diversity and dynamics;
- restore confidence in the system and enable citizens to participate in its development.

Based on this definition, what are the food system sustainability issues resulting from urbanization and industrialization?

Promoting Sustainable Agriculture to Feed Cities

Reclosing Fertilization Cycles and Reducing Pollution

Surplus food production is needed to feed nonfarming communities. In Mesopotanmia and Egypt in Ancient Times, and around the thirteenth century in the Khmer kingdom, cities were fed by agriculture that benefitted from fertilization via silt transfer from watersheds onto floodplains on both sides of the rivers. This is how Babylon (Van der Spek 2008), Alexandria (Viollet 2004) and Angkor (Evans et al. 2007) were able to grow. Note that this is how, for instance, Angkor managed to maintain a population of several hundreds of thousands of inhabitants, most of whom were involved in building temples.

That said, in 1800, only 3.4% of the world's population was urban, compared to about half in 2000! This soaring urbanization trend since the nineteenth century may be explained by changes in agricultural production methods that, in alternative ways, made it possible to generate surplus crops to feed a growing nonagricultural population. This trend generally occurred in two phases (Daviron 2016): first through an expansion of the cropping area and land colonization, and then by increased use of fossil coal as a fuel source. Coal enabled the development of transportation and provisioning via long-distance supply chains. Then the widespread use of petroleum prompted an increase in agricultural production by providing low-cost fuel to replace human and animal labour, with mechanization in turn leading to a marked increase in labour productivity. The use of petroleum also led to a decrease in material and energy extraction from forests, while also reducing areas devoted to producing animal feed—the main source of energy until then. These areas could subsequently be converted for food production. With the use of fossil fuels, it was no longer necessary to ensure cycling of nitrogen, which is essential for plant growth. At the expense of high fuel consumption (generally natural gas), the Haber-Bosh process made it possible to transform atmospheric nitrogen into chemical nitrogen fertilizer. Moreover, the use of mined phosphate and potassium overcame the need for faeces recycling, as was previously practiced (Box 1 and chapter "History of Urban Food Policy in Europe, from the Ancient City to the Industrial City").

Box 1: Urban Metabolism and Territorial Ecology

Sabine Barles and Gilles Billen

Territorial metabolism—with all of the caution required when considering the analogy to organic metabolism—refers to how territories consume and transform energy and matter, as well as how they mobilize and transform biosphere resources. This concept emerged from the idea that territories are dependent on these resources and modify the biosphere at different scales depending on how they use them. Interactions between societies and nature can be systematically analysed from this perspective: How much energy does a territory need to carry out all of its activities? How much material, i.e. water, foods and finished products, etc.? What happens to these flows once they have entered the territory since they have been used and transformed? In what form are they eventually returned to the environment? What are the impacts? Territorial ecology aims to place territorial metabolism in a spatial and social setting— the material and energy flows involved are the result of political, economic, social and technical choices, to name but a few. They reflect typical biosphere processes as well as the functioning of societies and cannot be analysed without taking this functioning into account. These flows link territories to others that supply them or receive their excreta, so a given territory's environmental footprint can be quite far from its limits—this situation is highly prevalent in cities, especially regarding their food.

Cities emerge since they offer at least some of their inhabitants the possibility of developing activities other than just subsistence production, including trade. Cities are thus models of social and spatial specialization. The upshot is that they outsource at least part (and now most) of their metabolism (Barles 2015). This outsourcing first concerned food flows and flows associated with heating and food preparation (referred to as 'energy' from the nineteenth century onward), while building materials were often extracted in the immediate vicinity of cities to facilitate transport. Outsourcing reached its peak with the industrial revolutions and urban transition—most cities have become entirely dependent on external territories for their supplies of ever more numerous materials and products, and for the elimination of their growing volumes of waste. Cities' energy and material consumption have substantially increased, while their metabolism has been levelling off—the reliance

(continued)

Box 1: (continued)

on fossil fuel (coal, hydrocarbons) and industrial fertilizers (atmospheric nitrogen fixed via the Haber-Bosch process, phosphate and potassium extracted from mines), etc., enabled human societies to disregard major bio-geochemical cycles and give up urban by-product recovery and recycling. The resulting environmental footprint is global, fragmented and deep-rooted.

Studies of material flows contributing to the urban food supply, especially in Paris which has been the focus of recent research (Abad 2002; Billen et al. 2011, 2014; Bognon 2014), have shown the extent to which these flows reflect the social relations and antagonistic trends that prevail in these cities. Until the early twentieth century, the Paris Basin (roughly the 200–250 km area around Paris) was the capital's main breadbasket hinterland, where the potential for producing agricultural surpluses was tailored to meeting the increased urban food demand induced by the considerable demographic growth rate. During the second half of the twentieth century, however, cropland specialization and disruption of the complementarity between crop and livestock farming, which was facilitated by the widespread application of industrial fertilizers, resulted in breakup of the supply area. The Paris Basin became devoted to export cereal crops and the *Grand Ouest* area to intensive livestock production based mainly on soybean imports from Latin America. The current Parisian paradox is that it relies on a quite locally circumscribed direct food market (within the limits of its specialization) whereas production is very globally oriented in the surrounding agricultural hinterland, with Paris now being just one of this hinterland's many market outlets. More than it might seem, the city has remained focused on its rural hinterland while the latter is very largely integrated in the globalization trend.

The future trajectory of the Parisian food system will depend on a tradeoff between two diametrically opposed trends, which correspond to two markedly different visions of the future of the city and agriculture. A centrifugal trend regarding the economic development of agriculture, agroindustries and supermarkets, which promotes the specialization and concentration of production and logistics tools in an increasingly globalized setting. In contrast, rising numbers of urban consumers, politically bolstered by local authorities, e.g. via the 2013 Master Plan for Ile-de-France Region, have created a centripetal trend with innovative forms of relocation of food links between the city and its historic hinterland.

(continued)

Box 1: (continued)

References

Abad R (2002) Le Grand Marché. L'approvisionnement alimentaire de Paris sous l'Ancien Régime. Fayard, Paris

Barles S (2015) The main characteristics of urban socio-ecological trajectories: Paris (France) from the 18th to the 20th Century. Ecol Econ 118: 177–185

Billen G, Barles S, Chatzimpiros P, Garnier J (2011) Grain, meat and vegetables to feed Paris: where did and do they come from? Localising Paris food supply areas from the eighteenth to the twenty-first century. Reg Environ Chang 12: 325–336

Billen G, Lasseletta L, Garnier J (2014) A biogeochemical view of the global agro-food system: Nitrogen flows associated with protein production, consumption and trade. Glob Food Sec 3: 209–219

Bognon S (2014) Les transformations de l'approvisionnement alimentaire dans la métropole parisienne. Trajectoire socio-écologique et construction de proximités. Thesis dissertation, Urbanism-Development, Université Paris 1

The now well known impacts of this industrialization of agricultural production are mixed. They are positive because this industrialization led to a sharp increase in food production, thus making it possible to feed a growing non-farmer population. On a global scale, one farm worker currently feeds around 5.5 people but, because of the still very high proportion of nonindustrialized agriculture in Asia and Africa, this ratio is over 1:140 in North America, according to FAOSTAT data.

The impacts are also negative because this production is achieved at the price of overconsumption of nonrenewable or slowly renewable resources: coal, oil and natural gas, mined phosphorus, as well as water due to the development of motorized irrigation. Fisheries resources have been exploited faster than their renewal capacity because of the motorization of fishing boats. In addition to this resource depletion, the environment is becoming saturated with pollutants: nitrogenous waste, GHGs, eutrophication and pesticides, leading to long-term environmental damage. Finally, agricultural specialization and intensive production strategies have contributed to the erosion of crop and livestock biodiversity, while also upsetting the biological balance (insects and microorganisms).

While it has been possible to feed fast growing cities so far, it seems that it will no longer be possible to continue to do so in the future if urbanization continues. Although industrial production methods could likely be streamlined to waste fewer resources and cause less pollution, they are still generally based on nonrenewable resource use, which is not sustainable in the long run. New production methods must be invented to be able to generate surpluses to feed cities but without using nonrenewable resources. Research on such alternative solutions has revealed that cities have a high resource supply potential, especially with regard to soil fertilization.

Until the early twentieth century, Shanghai and even Paris recovered and recycled faeces and urine produced by their inhabitants to fertilize the lands that fed them. Thereafter, the use of petroleum—hence abandoning the closure of nitrogen and phosphorus cycles—has resulted in almost unlimited extension of food supply areas of cities. Because of the reduction in transportation costs, cities get their food supplies from increasingly remote production areas. Urban metabolism analyses (Girardet 1999; Barles 2007; Billen et al. 2012) (Box 1) have shown that cities are now gigantic pumps that concentrate materials such as nitrogen and phosphorus. Food supplies come from areas much further away than rural areas where sludge from sewage treatment plants is dumped. These resources, part of which are discharged into the sea or rivers, have potential for use as fertilizer, but they are generally just wasted—becoming sources of pollution, especially nitrogen pollution (Lacroix 1995).

The reclosure of nitrogen, phosphorus and more generally biomass cycles is therefore a key urban food policy challenge. It highlights that cities can help promote new more sustainable modes of agricultural production. In the interstices of residential neighbourhoods, in periurban areas with high land constraints, so-called 'urban-farmers' practicing urban agriculture also invent new forms of production that are little or not at all reliant on chemical pesticides, while focusing on crop associations, saving water and making effective use of waste. The expertise that prevails in these nonconventional agricultural systems deserves to be studied and eventually developed.

This first issue highlights that urban food policies do not solely concern urban populations, but they can help shape agricultural production models to make them more sustainable.

Promoting Employment-Generating Agricultural and Agrifood Models

Industrialization—by markedly boosting labour productivity via mechanization—frees up the workforce in sectors affected by this process. In European countries, the transfer of labour from primary to secondary and tertiary sectors has been ongoing for almost two centuries. The rates of labour productivity growth and of freeing up the agricultural workforce have been relatively consistent with the pace of job creation in industrial and service sectors.

The situation is different in more recently industrialized countries. With urbanization, the emergence of the agrifood sector initially involves a surge of activity and job creation, especially for women. Subsequently, industrial mergers and foreign investment by large capital-intensive agroindustrial groups have the opposite effect. They compete with and rapidly eliminate thousands of micro-businesses and

SMEs (Malassis and Padilla 1986), with the risk of creating situations of high income inequality and mass unemployment (Weatherspoon and Reardon 2003; Cadilhon et al. 2006; Grain 2014). But recently industrialized countries are also in full demographic transition. It is estimated that in sub-Saharan Africa, given the extent of population growth, around 30,000 jobs per million inhabitants and per year should be created (Beaujeu et al. 2011). The agricultural, agrifood processing, distribution and catering sectors are emerging as the main potential sources of employment to meet this challenge (Losch et al. 2012; Farm 2016).

Through their potential to promote the development of relatively labour-intensive businesses—e.g. via commercial urbanism or job creation support—cities have the power to influence production models, especially in the agrifood sector.

Reducing Reliance on Remote Supplies and Reconnecting Cities to Their Hinterlands

Dependence on International Markets and Risks Associated with Market Volatility

Urbanization is considered by many observers as a key factor of international market food dependence (Pingali 2007; Wilkinson 2008; Porkka et al. 2013). Some authors consider that this dependence results from the domination of industrialized countries. Dumping of agricultural exports practiced by the United States and Europe until the 1990s, compounded by food aid, led to a competitive situation with regard to local production (Barrett and Maxwell 2007). These regular imports got city dwellers used to consuming foreign products even though they differed from those produced locally. For others, such as Abdou Touré (1981) and Victor Scardigli (1983), the economic and political dominance of Western countries prompts urban consumers to imitate their models. Although these theories of behavioural mimicry and extroversion have been criticized on the basis of empirical findings (Odeyé and Bricas 1985; Requier-Desjardins 1989, on food in Africa; Appadurai 1996, more generally), they implicitly underpin other theories on the 'westernization' of food (see studies of Popkin, Pingali, Usitalo, Goodland, etc.), the 'Coca-colonization' (Webster 1989) and 'McDonaldization' (Ritzer 2011) of diets.

Other authors such as Olivier Sudrie (1985) further explain import dependence as a strategic option for countries to cost-effectively secure food supplies for their potentially politically unstable urban population. This analysis is in line with the historical interpretation of Fernand Braudel (1979) concerning urbanization between the fifteenth and eighteenth centuries, while pointing out the difficulties that prevailed in building domestic markets: "A relatively flourishing external market usually preceded the laborious unification of the national market" (Braudel 1979: 332).

In other words, it has often been easier for countries to procure distant food supplies while building their national markets. More recently, in an economic liberalization setting, the period of low price stability that prevailed in international markets between the 1980s and the late 2000s prompted many countries to continue securing their supplies via imports.

The soaring international market prices and resulting urban riots that took place in 2008 and 2010 revealed the vulnerability of this food security option. International markets, which operate under a strongly finance-driven tight flow regime with low stocks, are now considered more volatile and riskier than before. In recent years, there has been a noticeable trend towards recapturing urban markets via local production supplies to reduce the dependence on imports. The challenge is not solely to reduce price volatility risks by diversifying supply sources. Indeed, reconnecting cities to their hinterlands is a further way to generate jobs and income in rural areas, thus avoiding rapid outmigration to cities.

Cities as Drivers of Agricultural Production

The focus on cereals in discussions regarding food dependency should not mask the fact that cities are already connected to their hinterlands. Various authors consider that urbanization has led to domestic market growth to the benefit of farmers. So-called food crops, initially considered essentially as being destined for on-farm consumption, became cash crops mainly to supply urban markets (see studies of Bricas, Chaleard, Moustier, Tacoli, etc.). In addition to being a driving force behind agricultural development, urbanization generates a range of intermediation activities with the rural community, promoting the diversification of income sources in rural areas and job creation in cities (Lopez and Muchnik 1997). A recent analysis of surveys of food consumption patterns in West Africa—a region that has experienced a sharp increase in food imports since the 1960s—revealed cities' high dependence on imported rice and wheat. Imports of these two cereals represent roughly two-thirds of the value of starch-based commodities consumed in cities. However these starchy products themselves account for only about a third of the economic value of urban household food consumption. The rest, i.e. animal products (one third), sauce products (legumes, vegetables, oil and condiments), fruit and sweet products (including drinks) (for the last third), are mostly produced locally (Bricas et al. 2016).

It is therefore not inevitable that food globalization will fuel urbanization, but the role of local production as an urban market driver is still a major policy issue that cities can help address. In this respect, some rely on infrastructure (e.g. farmers' markets or wholesale markets), on regulation or contracting, especially for public procurement (e.g. for collective catering), on local product promotional operations, or on investment in rural areas or financial support (e.g. to rural agroindustries), thus helping to boost the added value for farmers.

Rebalancing City/Rural Relationships

This connection nevertheless does not represent a balance of power between cities and rural areas (Lipton 1977). Cities—since they are hubs where trade flows converge—force competition between food production areas, thereby putting pressure on farmers. Urban markets dictate prices for local products that farmers consider unprofitable, as well as hard to meet product quality requirements. This pressure can lead to paradoxical situations whereby family farmers generate production surpluses to feed cities while experiencing food insecurity because of their low incomes. In this regard, farmers living far from cities are often obliged to rely on intermediaries to gain access to urban markets, which in turn further reduces their income (Tacoli 1998).

Rural poverty and widening gaps regarding living standards between cities and rural areas often trigger rural outmigration, in turn raising problems for urban authorities. The rebalancing of these city/rural interactions is a rising concern for cities and their inhabitants. This may prompt investment in rural areas to help maintain rural communities, which is a priority of some associations of formerly rural people who have taken up residence in cities (Ouattara 2005) and some municipalities (e.g. Hanoi, with rural development contracts in Vietnamese provinces, or Medellin, with support given to rural agroindustries to increase the added value for Colombian farmers). This rebalancing may also be in the form of contractual agreements between collective caterers and rural production areas to ensure supplies of quality products and better financial gains for farmers.

A movement is emerging that is opposed to the idea of rural areas being reduced to the role of nurturing cities and is striving to create new more balanced and sustainable city/rural relationships. That is the goal of the city-region food systems concept, which stresses the benefits of relocating agriculture to areas around the markets served, and with better balancing of relationships between these complementary worlds (Jennings et al. 2015; Dubbelling et al. 2015).

Facilitating the Movement of Goods and Access to Food

Reducing Last-Kilometre Transportation Costs

Daily feeding of large population concentrations raises substantial logistics problems. Transportation, storage and distribution result in high flows of goods in constrained spaces (Gaigné et al. 2011). Most initial urban markets are built at the city core. The market is a place of exchange which gives life to the city and many other activities develop in the vicinity of these markets. As the city expands this central market model is generally replicated in the new neighbourhoods, until the point when such expansion of the urban area and building density starts to hamper the flow of wholesale goods. Many cities are then forced to geographically separate wholesale and retail functions, with wholesale infrastructure being set up at the outskirts of cities to decongest the city of supply trucks (Densley and Sanchez-Monjo 1999). Moreover, sanitary constraints, including the management of livestock being transported into cities, also contributes to the reorganization of flows. Centralized slaughterhouses are generally built at the city gateways, which avoids livestock entering the neighbourhoods and enables centralized animal health control (Fitzgerald 2010).

This logistics restructuring generally began taking place during the twentieth century and is still ongoing without any real energy constraints. The environmental impact of the different logistics models, particularly the issue of transportation energy costs, has become problematic with the relatively recent threat of soaring petroleum prices. However, few studies to date have focused on this issue (Masson and Petiot 2012). In Great Britain, Alison Smith et al. (2005) calculated that household car food shopping trips—mainly to supermarkets located on the outskirts of cities—represented 47.5% of the overall food transport distance covered in 2002 and 13% of CO_2 emissions and 40% of total social costs due to food transport. At a time when new potential logistics models are emerging, along with the development of e-commerce and new delivery modes (click & drive, drone), the last-kilometre organization issue has become important from an environmental standpoint. Sally Cairns (2005) thus showed that home food delivery, rather than household food shopping trips, could reduce the overall food transport distance covered by 70% (Box 2). Reconfiguring urban food logistics is also a social issue (Morana and Gonzalez-Feliu 2011). The food supply mode also impacts consumers' relationship with food. Depending on whether or not there is human contact in the commercial exchange, this mode more or less leads to commodity standardization, while also having an impact on the diversity of the supply.

Preventing the Emergence of Food Deserts

The development of supermarkets can generate competition with local shops and even eliminate them. This often results in a dearth of grocery stores in some city neighbourhoods, which means that only households with medium-distance means of transport will have access to healthy and nutritionally suitable food (Beaulac et al. 2009). Taking the nutritional impacts of these so-called 'food deserts' into account has thus become a strategic priority in commercial planning policies in some cities in the Global North.

This fight to prevent the development of food deserts nevertheless does not always involve preserving small shops in which products may be sold at relatively high prices. In some cities, such as Medellin in Colombia, a municipal policy to ensure that the poorest people will have access to quality food has led to the building of central purchasing units in popular neighbourhoods of the city, where only small shops previously prevailed (Correa Peläez 2016).

Box 2: Who Generates Urban Conurbation Food Supply Flows?[1]

Jean-Louis Routhier

Under the conventional pattern that prevailed until the early 1960s, households mainly shopped on foot at local stores that were provisioned by truck. However, it is now possible, thanks to the widespread adoption of car transport, for consumers to fetch supplies at stores located far from their homes. The pattern has thus evolved in favour of the development of supermarket food shopping by car. Supermarkets in turn have a role as urban distribution warehouses, providing an alternative to operators who provision local shops. The supply is therefore split between shops and households.

Provisioning Food Shops

In the light of the findings of the ETMV 'goods in cities' survey conducted in Ile-de-France region in 2011, if we just focus on the store supply chain, it emerges that out of the 4,250,000 goods deliveries or pick-ups that take place weekly, 850,000 (19%) concern foodstuffs. In detail, these movements are split into three parts:

- a large third upstream share related to industrial and small-scale production (11%), wholesalers (20%) and warehouses (6%);
- around 45% related to catering for cafés, hotels and restaurants (31%), and corporate catering (13%);
- finally, retailers represent barely 20%, headed by small traditional food businesses (groceries, bakeries, butcheries and markets; 13%), followed by supermarkets (only 5%; but obviously in batches of several tonnes and delivered by heavy trucks!).

Light vehicles are mostly used to carry out these operations (54%), compared to 39% trucks of more than 3.5 t and 7% semi-trailers.

The weight and packaging of deliveries of course differ from one end of the chain to the other. On average, the mean tonnage per delivery to warehouses, industries and hypermarkets ranges from 4 to 7 t, while for supermarkets and wholesalers it is 2.5 t, and 1 t for small supermarkets. Finally, cafés, groceries and markets receive batches of around 500 kg. For home deliveries of foodstuffs, the mean weight ranges from 50 to 90 kg, whereas when individuals themselves transport their purchases, the foodstuff weight ranges from a few kilogrammes when they are walking to 20 kg on average when they are in their private cars.

(continued)

[1] I would like to thank Florence Toilier, Marc Serouge and Mathieu Gardrat for their invaluable help in generating several of the results summarized here.

Box 2: (continued)

Consumer Shopping Trips

The EGT 'general transport' survey conducted in Ile-de-France in 2010 revealed that residents of this region make 34 million shopping trips a week, 33% of which take place on the weekend. Overall, 42% are by car, 50% walking, 8% by public transport or soft modes. With a corrected car occupancy rate of 1.29 passengers, there are 11 million purchase pick-ups a week in private cars. Foodstuffs should be separated from the purchases in order to be able to compare these findings with the figures in the previous paragraph, but this is currently hard to accomplish because daily mobility surveys generally do not specify the type of goods purchased. Moreover, since it is not possible to specifically focus on food purchases, we compare household shopping trips overall and movements of goods generated by the business supply chain (consumer goods producers, wholesalers and retailers). This results in 1,700,000 goods deliveries and pick-ups a week, which is about sixfold fewer than private car shopping trips.

The most relevant measure for assessing the sustainability of urban food flows is, however, not the number of trips. It is better to focus on road occupancy by motor vehicle traffic to account for the scarcity of road space that typifies the urban space. The ETMV and EGT surveys make it possible to estimate this road occupancy linked to supplying businesses and households.

80 million km are travelled weekly by Ile-de-France residents during shopping trips, 70% of which is to supermarkets. Compared to the 41 million km-PCE[2] generated by the retail supply chain in 1 week, the road occupation for shopping trips is twofold greater than that of vehicles involved in the business supply chain.

These few figures clearly show that urban food supply issues are very closely related to consumer mobility practices for shopping.

Are 'New Practices' Exemplary?

In this domain, as elsewhere, there is marked diversification in food supply modes in the cities: Markets, local shops, supermarkets, as well as the many cafés and restaurants. The practices are also diversified, ranging from 'click & drives' to 'relay point' (pick-up and drop-off location) and home deliveries, as well as neighbourhood AMAPs (consumer associations providing bulk

(continued)

---

[2] PCE: passenger car equivalent: a weighting factor used to take the vehicle road footprint into account (a light commercial vehicle [LCV] = 1.5 PC; a straight truck = 2 PC; a heavy articulated truck = 2.5 PC). Calculations performed using the FRETURB© model.

Box 2: (continued)

delivery of produce) and even harvesting of produce in market gardens by nature-loving urban households. The supply pattern is therefore further evolving and becoming more complex.

Preliminary calculations based on a survey conducted in late 2015 in metropolitan Lyon (France) showed that each household generates about 20 equipment acts of purchase a year that are 'disconnected' with shopping trips, i.e. receipt of goods by the buyer does not take place at the same time as the act of purchase: Home or relay point delivery following a phone or online purchase, etc. if this figure is transposed to conventional shopping trips in the Ile-de-France area, it would represent around 6% of the total shopping trips by individuals and 18% of their car shopping trips. Of these 20 disconnected annual purchases, only four involved food shopping and three online meal orders. This emerging and fast-growing phenomenon has yet to structurally modify flows and road occupancy. These new practices are likely to add to, or partly or totally supplant, conventional shopping trips. Ongoing research should make it possible to assess the impacts of these new behavioural and operational trends on the city.

This text is based on data processed by the *Laboratoire aménagement économie transports* (LAET) research unit (Lyon) based on findings from the following reference sources.

References

'Deferred Household Purchases' survey, 2015, Metropolitan Lyon – ADEME
EGT 2010 survey, Ile-de-France region, DRIEA, IAU, STIF
FRETURB© model applied to Ile-de-France, DRIEA-LET
TMV 2011 surveys, Ile-de-France Region, ADEME, DRI of the French Ministry of Transport

Promoting More Sustainable Urban Food Styles

To Cope with Food Style Changes Related to Urbanization…

The growing remoteness of agricultural production areas that supply cities leads to the development of processing activities to reduce the volume and weight of food to be transported, while extending their shelf life: drying, shelling and extraction of the useful fractions (oil, sugar, juice, etc.). The agrifood sector is also developing to provide urban dwellers already processed foods that are more convenient to use and increasingly diversified (Colonna et al. 2011). Human concentration also leads to the development of distribution and catering activities in cities. These activities, which are initially carried out on a small scale, have become industrialized with the gradual increase in consumer purchasing power, thus providing income for an increasing number of services incorporated in food commodities, i.e. products that are precooked, have a long shelf life or are divided into individual portions. These processed commodities are promoted by agrifood companies, sometimes with a lot of publicity (Kearney 2010; Monteiro and Cannon 2012; Stuckler and Nestle 2012).

The higher purchasing power of urban dwellers generally results in an increase in the consumption of animal products, i.e. meat in many countries and dairy products in places such as India.

...the Environmental Cost of Diets Has to be Reduced

From an environmental viewpoint, this dietary shift towards more industrially processed products is reflected—as for agricultural production—by high nonrenewable or slowly renewable resource consumption and pollution. The extent of the agrifood sector's contribution to environmental degradation is hard to calculate due to the absence of data differentiating the food contribution from industrial, transportation and other factors. The NGO Grain (2011) attempted this calculation on a global scale and concluded that the food processing sector, including refrigeration and packaging, accounts for 13–15% of total GHG emissions, with transport only representing 5–6%, distribution 1–2%, and agricultural organic and food waste decomposition 3–4%. Overall, the agrifood sector represents 15–20% of total GHG emissions, compared to 11–15% for agricultural production alone and 15–18% for land-use changes, mainly related to deforestation, 70–90% of which is caused by farmland expansion. When considering environmental impacts other than GHG emissions (acidification, ecotoxicity, eutrophication, etc.), the contribution of the food sector seems to be lower. Food and beverage products contribute 20–30% to the overall environmental impact (Tukker and Jansen 2006).

Packaging pollution is also a noteworthy issue. The proliferation of individual packaging, ranging from simple polyethylene bags to more elaborate packages combining plastics, aluminium and cardboard, is a major source of pollution and concern in cities where such packaging occurs in dense concentrations (Marsh and Bugusu 2007). This has prompted some cities (e.g. Dacca, San Francisco, Kinshasa and Dakar) to regulate the use of certain packaging, thus forcing companies to seek alternative solutions.

From a biodiversity standpoint, the standardization of products and the growing need for uniform quality raw materials because of technological processing constraints leads to erosion in the diversity of raw material varieties used (McKinney 2006). Hence the number of varieties that may be used by the agrifood industry is declining. Nevertheless, a very diverse range of processed products could be obtained via raw material fractionation (Soler et al. 2011). Through their preferred logistics models, cities have the levers needed to help preserve biodiversity. The development of central purchasing centres serving supermarket chains results a reduction in the number of items marketed due to volume bulking. In comparison, the wholesale market model promotes the marketing of a wider variety of products as it provides a better linkage between smallholder farmers and niche markets (De Raymond 2010).

The industrialization of food products and the increased consumption of animal products decreases the energy efficiency of food. A growing amount of energy is required for animal production, food processing, packaging and marketing to be able to guarantee a daily intake of 2000–2500 kcal per person (Pimentel and Pimentel 2007). The worldwide adoption of the so-called Western consumption model is turning out to be unbearable for global resources, while also generating high levels of pollution. The impacts of this trend, however, are not solely environmental in scope.

Jointly Combating Obesity and Micronutrient Malnutrition

Changes in diet related to urbanization also have nutritional impacts. In rural areas, food insecurity has long been linked (and continues to be in some regions) to food supply insufficiency because of the low or highly variable production conditions, high postharvest losses or limited market access due to remoteness. Difficulties in eating properly in urban areas are primarily related to economic poverty rather than to the available food supply. In developing countries that have undergone rapid urbanization, many urban inhabitants live below the poverty line as they do not have a sufficiently lucrative job to meet their baseline needs. According to World Bank poverty statistics, this situation prevailed in 50% of the urban population in Mexico in 2014, 36% in Côte d'Ivoire in 2015, 25% in Colombia in 2014 and 21% in India in 2009. These people do not have enough purchasing power to gain access to foods of good nutritional and health quality, and they are very vulnerable to price hikes.

Middle classes are also emerging in cities of the Global South despite the extent of urban poverty. The average living standards of urban households are everywhere higher than those of rural households (Ginneken 1976). These cities are thus the main locus of the so-called double nutritional burden, i.e. the coexistence of undernutrition and overnutrition (Boutayeb 2006). Undernutrition due to insufficient calorie or protein intake is not as prevalent as it is in rural areas. This condition is especially the result of micronutrient deficiency (iron, vitamin A, zinc, etc.) which causes stunted growth. But one specific feature of cities is the considerable increase in diseases related to overconsumption of fat, sweet and salty products, combined with reduced physical activity, especially amongst the emerging middle class: overweight and obesity, often associated with diabetes, cardiovascular diseases and certain cancers (Popkin 1999; Maire and Delpeuch 2004; Goryakin and Suhrcke 2014).

After long having favoured options to boost awareness and educate populations, policies are now more oriented towards actions that will enhance consumers' knowledge and attitudes as well as their environment (Cohen and Ilieva 2015). The food supply is no longer seen as being an independent response to a demand but rather as a means to shape it (Lahlou 2005). Consumers' behaviours are not solely shaped by their knowledge and wishes, the facilities and incentives of their environment or foodscape also create behavioural habits. This major shift in perspective highlights the role that urban policy could potentially have, e.g. by the way collective catering is organized, as a means of learning about food habits, or by commercial urbanism, e.g. spaces left for gardens, markets, restaurants, waste management organization, etc.

Enhancing the Health Quality of Food Consumed by Poor Communities

In poor city neighbourhoods, work schedule and traveling hour constraints as well as the lack of space in precarious habitats (Satterthwaite et al. 2010) often leads to a reliance on popular restaurants and street food. These small-scale activities generate job opportunities but are carried out with limited resources, i.e. scant access to drinking water or quality raw materials, unsanitary environments, lack of food storage facilities, etc. (Henson 2003; Broutin and Bricas 2006). Moreover, anonymity and low institutional quality control capabilities promote unfair commercial exchange behaviour. The informal food sector—although providing essential services to feed populations with limited purchasing power—is thus often considered as a generator of consumer health risks (Ekanem 1998; Winarno and Allain 1991).

Food insecurity in these environments also involves difficult access to healthy food. More generally, market segmentation takes place with a supply of strict quality controlled products targeting relatively rich consumers, alongside the development of a supply targeted to populations with low purchasing power. This supply comes mainly from informal microenterprises and itinerant trade, but also from more specific sectors in which unsold products from formal markets are recycled for popular markets.

The development of market stalls to improve the health conditions, the creation of equipped areas devoted to popular restaurants and the training of craftspeople, are different ways that some cities implement to enhance the health quality of food.

Reducing Inequality and Power Asymmetry

In addition to the essential rebalancing of power between small and large enterprises and between cities and rural areas, reducing inequality within the urban population is a major social and political challenge.

The concentration of poor people in cities who have no access to the abundant available supply, while living alongside rich and super rich populations, can generate a powder keg of social instability in crisis situations, as shown by the urban riots that took place in 2008 and 2011 triggered by the rising energy and food prices. Many of these clashes resulted in the destruction of public infrastructure, violence and the fall of governments. They revealed the importance of food price stabilization (HLPE 2011), as well as the employment issue.

Beyond this political risk, geographic, economic, cognitive and political distancing can create a credibility gap between consumers and their food system, further accentuated by power asymmetries.

This distancing results in a feeling amongst consumers that they are losing control of the system—a decline that is spurred by the acceleration of modernity (Rosa 2010)—thus lowering the risk acceptability (Slovic 1987). Personal confidence in food quality is less dependent on each person's sensory and cognitive cues and relies more on third parties (intermediaries, labels, prices, brand reputation and State seals of approval) (Cheyns and Bricas 2003). This process is not self-evident and may give rise to worry and suspicion, especially when these third parties prove to be fallible (mad cow crisis, contaminated blood crisis and Chernobyl cloud crisis) or are suspected of protecting their own economic interests before the health of consumers.

The growing anxiety amongst food consumers is further accentuated by the individualization of their eating habits (Fischler and Masson 2008). What was once 'self-evident', a routine embedded in a universe of rules and conventions, has become a question of individual choice. In the abundant and diverse supply, food consumers must now make choices between nutritional, taste, environmental, practical and price attributes, etc., but without having any solid benchmarks in the face of controversies and uncertainties about food quality. The increase in everyone's freedom of choice has led to a paradox (Schwartz 2004), i.e. it generates anxiety and a feeling of not being able to properly orient one's diet (Poulain 2002; Poulain and Corbeau 2002). Finally, consumer over-responsibility tends to clear suppliers of their responsibility to shape consumption patterns, in turn accentuating consumer anxiety (Figuié and Bricas 2011).

The search for new proximities may be viewed as a reaction to this distancing and individualization, while reassuring food consumers. Many urban food policies, especially in the most industrialized countries, are focused on relocating supplies and building more inclusive governance. Local supplies, short food supply chains, direct marketing, territorial food systems, local food policy councils, etc., are amongst potential ways to reassure food consumers, while giving them the feeling that they are recovering some control over their diets.

Living Together

The city is a space where populations of various cultural origins and contrasting social backgrounds coexist. For some, it is a place of deculturation and anomy, generating social tension or even violence exacerbated by economic precarity and growing inequality (Hérault and Adesanmi 1997). In the food domain, the city has often been considered as a place where people lose their 'traditional' bearings that serve to structure food habits (Fischler 1979; Mestdag 2005), and as a privileged place for the extension of global culture (Westernization). In response, these phenomena may lead to demands for respect of the specific food habits of some cultural or religious groups—a sign of hyperculturation (Barber 2010; Jourdan and Riley 2013).

Other authors consider instead that the city is a space for integration and development of a unique culture. Urban identity is formed especially via cooking, i.e. through gastronomy and chefs associated with the city or, more often, by the development of popular kitchens associated with small-scale catering (Bricas and Odéyé 1985). In Africa, Dakar-style fish and rice (*thiébou diène*) (Sankale et al. 1980), Abidjan-style cassava semolina (*garba d'attiéké*) (Konaté 2005) and Ouagadougou-style cornmeal and greens (*baabenda*) are typical urban dishes that contribute to urban cultural integration and identity building.

In one way or another, changes in food styles associated with urbanization are major sociocultural issues, including food heritage changes in a setting of strong external influences, and living together in a setting of growing social tension (Tibère 2009).

Here again cities can, through their food policies, help inhabitants live better together. Cooking and gastronomy are eminent cultural and identity enhancing activities and can contribute to generating inclusive identities, understanding and appreciating other cultures, for instance via collective catering.

Conclusion

Literature on sustainable urban food systems often tends to be focused on the relationship of cities with agriculture, particularly on the food supply issue. These studies are part of a movement seeking new geographical (promotion of local commodities), economic (short supply chains), cognitive (learning about agriculture through community gardening) and political (food governance) proximity. The city region food systems (CRFS) concept promoted by several international organizations and NGOs also reflects this movement. Other studies with the same rationale of bringing urban and rural areas closer together are also focused on the possible closure of cycles of materials that tend to build up in cities and which could be effectively used in agriculture.

These studies place less emphasis on the internal organization of the food system within cities, on the effects of forms of urbanization or on city dwellers' behaviour. However, the non-sustainability of urban food systems is also a result of urban lifestyles, spatial configurations and consumption practices. In this sense, all food system sustainability issues related to urbanization cannot simply be solved by reconnecting cities to their rural environment.

Cities have long had resources that impact food. They manage land and can thus maintain agricultural activities within or around the city. They manage the space and thus shape the foodscapes and dictate the location of shops and market places. They organize their supply logistics through wholesale market management, with a direct impact on the type of agriculture that these modes of management promote. They often manage collective catering, particularly for children and youth—a multifunctional tool for food and nutritional security, education and social integration. They encourage the development of economic activities, especially in the agrifood sector, by developing spaces devoted to the creation or establishment of businesses, while taking advantage of taxation conditions and organizational services. They manage waste, which in turn is a resource that may be useful in agriculture.

These many levers, which generally concern distinct sectors, are now being reconsidered from a more integrative standpoint. It is no longer a question of guaranteeing a constant supply, of ensuring universal access to quality food or of reducing congestion in cities. Increasingly, these levers are now used to help invent new and more sustainable food systems.

References

Appadurai A (1996) Modernity at large: cultural dimensions of globalization. University of Minnesota Press, Minneapolis

Barber BR (2010) Jihad vs McWorld. Random House, New York

Barles S (2007) Feeding the city: food consumption and flow of nitrogen, Paris, 1801–1914. Sci Total Environ 375(1):48–58

Barrett CB, Maxwell D (2007) Food aid after fifty years: recasting its role. Routledge, London, 314 p

Beaujeu R, Kolie M, Sempéré JF, Uhder C (2011) Transition démographique et emploi en Afrique subsaharienne. Comment remettre l'emploi au cœur des politiques de développement, À savoir, vol 5. Ministère des affaires étrangères et européennes, AFD, Coll, Paris, p 213

Beaulac J, Kristjansson E, Cummins S (2009) A systematic review of food deserts, 1966–2007. Prev Chronic Dis 6(3):A105

Billen G, Garnier J, Barles S (2012) History of the urban environmental imprint: introduction to a multidisciplinary approach to the long-term relationships between Western cities and their hinterland. Reg Environ Chang 12(2):249–253

Boutayeb A (2006) The double burden of communicable and non-communicable diseases in developing countries. Trans R Soc Trop Med Hyg 100(3):191–199

Braudel F (1979) Civilisation matérielle, économie et capitalisme, xvᵉ–xviiiᵉ siècle. Tome 3: le temps du monde. Armand Colin, Paris

Bricas N, Odéyé M (1985) À propos de l'évolution des styles alimentaires à Dakar. In: Bricas N, Courade G, Coussy J, Hugon P, Muchnik J (eds) Nourrir les villes en Afrique sub-saharienne, Coll. Villes et entreprises. L'Harmattan, Paris, pp 179–195

Bricas N, Lamine C, Casabianca F (2013) Agricultures et alimentations : des relations à repenser ? Nat Sci Sociétés 21(1):66–70

Bricas N, Tchamda C, Mouton F (eds) (2016) L'Afrique à la conquête de son marché alimentaire intérieur. Enseignements de dix ans d'enquêtes auprès des ménages d'Afrique de l'Ouest, du Cameroun et du Tchad, Coll. Études de l'AFD, Cirad, Afristat. AFD, Paris, p 130

Broutin C, Bricas N (2006) Agroalimentaire et lutte contre la pauvreté en Afrique subsaharienne; le rôle des micro et petites entreprises. Éd. du Gret, Paris, 128 p

Burlingame B, Dernini S (Eds) (2010) Sustainable diets and biodiversity. Directions and solutions for policy, research and action. In: Proceedings of the international scientific symposium biodiversity and sustainable diets united against hunger, FAO Headquarters, Rome, 3–5 November 2010, p 309

Cadilhon JJ, Fearne AP, Figuie M, Giac Tam PT, Moustier P, Poole ND (2006) The economic impact of supermarket growth in Vietnamese food supply chains. Int J Environ Cult Econ Soc Sustain 2(7):1–12

Cairns S (2005) Delivering supermarket shopping: more or less traffic? Transp Rev 25(1):51–84

Cheyns E, Bricas N (2003) La construction sociale de la qualité des produits alimentaires ; le cas du soumbala, des céréales et des viandes sur le marché de Ouagadougou au Burkina. Montpellier, Cirad, Série Alimentation, Savoir-faire et Innovations en agroalimentaire en Afrique de l'Ouest, p 82

Cohen N, Ilieva RT (2015) Transitioning the food system: a strategic practice management approach for cities. Environ Innov Soc Trans 17:199–217

Colonna P, Fournier S, Touzard JM (2011) Systèmes alimentaires. In: Esnouf C, Russel M, Bricas N (eds) Pour une alimentation durable. Réflexion stratégique DuALIne. Quæ, Paris, pp 59–84

Correa PF (2016) Alianza por el buen vivir. Fortalecimiento de Sistemas Agroalimentarios. 43rd Committee on world food security, forum on urbanization, rural transformation and implications for food security and nutrition, Rome, 20 October 2016

Cour JM (2004) Peuplement, urbanisation et transformation de l'agriculture : un cadre d'analyse démo-économique et spatial. Cah Agric 13(1):158–165

Daviron B (2016) Agriculture et économie : du solaire au minier… et retour ? Agron Environ Soc 6(1):23–34

De Raymond AB (2010) Dispositifs d'intermédiation marchande et politique des marchés. La modernisation du marché des fruits et légumes en France, 1950–1980. Sociologie du travail 52(1):1–20

Densley B, Sanchez-Monjo E (1999) Wholesale market management. A manual. FAO Agric Serv Bull 140, 96 p

Dubbeling M, Renting H, Hoekstra F, Wiskerke JSC, Carey J (2015) City region food systems. Urban Agriculture Magazine, p 29

Ekanem EO (1998) The street food trade in Africa: safety and socio-environmental issues. Food Control 9(4):211–215

Evans D, Pottier C, Fletcher R, Hensley S, Tapley I, Milne A, Barbetti M (2007) A comprehensive archaeological map of the world's largest preindustrial settlement complex at Angkor, Cambodia. Proc Natl Acad Sci 104(36):14277–14282

Farm (2016) Emploi en Afrique: et si l'agroalimentaire était une solution ? International conference, 8 December 2016. http://www.fondation-farm.org/spip.php?page=article&id_article=992

Figuié M, Bricas N (2011) Réinvestir la régulation publique. Le risque d'une sur-responsabilisation des consommateurs. Problèmes politiques et sociaux 982:99–100

Fischler C (1979) Gastro-nomie et gastro-anomie. Communications 31(1):189–210

Fischler C (1990) L'homnivore. Le goût, la cuisine et le corps. Odile Jacob, Paris

Fischler C, Masson E (2008) Manger : Français, Européens et Américains face à l'alimentation. Odile Jacob, Paris

Fitzgerald AJ (2010) A social history of the slaughterhouse: from inception to contemporary implications. Hum Ecol Rev 17(1):58–69

Gaigné C, Capt D, Faguer E, Frappier L, Hilal M, Hovelaque V, Le Cotty T, Parrot L, Schmitt B, Soulard C (2011) Urbanisation et durabilité des systèmes alimentaires. In: Esnouf C, Russel M, Bricas N (eds) Pour une alimentation durable. Réflexion stratégique DuALIne. Quæ, Paris, pp 123–142

Ginneken WV (1976) Rural and urban income inequalities in Indonesia, Mexico, Pakistan, Tanzania, and Tunisia. International Labour Office, Geneva, 67 p

Girardet H (1999) Sustainable cities: a contradiction in terms? In: Satterthwaite D (ed) The Earthscan reader in sustainable cities. Earthcan, London, pp 413–425

Godard O (1994) Le développement durable. Paysage intellectuel. Nat Sci Sociétés 2(4):309–322

Goryakin Y, Suhrcke M (2014) Economic development, urbanization, technological change and overweight: what do we learn from 244 demographic and health surveys? Econ Hum Biol 14:109–127

Grain (2011) Food and climate change. The forgotten link. Grain, Barcelona, 5 p

Grain (2014) Food sovereignty for sale: supermarkets are undermining people's control over food and farming in Asia. Grain, Barcelona, 19 p

Henson S (2003) The economics of food safety in developing countries. ESA Working Paper 19(3): 19–30

Hérault G, Adesanmi P (Eds) (1997) Jeunes, culture de la rue et violence urbaine en Afrique. In: Actes du symposium international d'Abidjan. Ibadan, IFRA, 5–7 mai 1997, p 419

HLPE (2011) Price volatility and food security. A report by the High Level Panel of Experts on Food Security and Nutrition of the Committee on World Food Security, Rome, CFS, p 79

IPES Food (2015) The new science of sustainable food systems. Overcoming barriers to food system reform. IPES Food, Brussels, 21 p

Jennings S, Cottee J, Curtis T, Miller S (2015) Food in an urbanized world. The role of City region food systems in resilience and sustainable development. International Sustainability Unit, London, 92 p

Jourdan C, Riley KC (2013) Présentation: La glocalisation alimentaire. Anthropol Soc 37(2):9–25

Kearney J (2010) Food consumption trends and drivers. Philos Trans R Soc London B: Biol Sci 365(1554):2793–2807

Konaté Y (2005) Abidjan : malentendu, poésies et lieux propres. Outre-Terre 11(2):319–328

Krausmann F, Haberl H (2002) The process of industrialization from the perspective of energetic metabolism: socioeconomic energy flows in Austria 1830–1995. Ecol Econ 41(2):177–201

Lacroix A (1995) Des solutions agronomiques à la pollution azotée. Cah Agric 4(5):333–342

Lahlou S (2005) Peut-on changer les comportements alimentaires? Cahiers de nutrition et de diététique 40(2):91–96

Lamizet B (2007) La polyphonie urbaine : essai de définition. Communication et organisation 32:14–25

Lewin RA (2009) Merde: excursions in scientific, cultural, and socio-historical coprology. Random House, New York

Lipton M (1977) Why poor people stay poor: urban bias in world development. Econ J 87(347):611–613

Lopez E, Muchnik J (1997) Petites entreprises et grands enjeux. L'Harmattan, Paris

Losch B, Fréguin-Gresh S, White ET (2012) Structural transformation and rural change revisited: challenges for late developing countries in a globalizing world. World Bank Publications, Washington, DC

Maire B, Delpeuch F (2004) La transition nutritionnelle, l'alimentation et les villes dans les pays en développement. Cah Agric 13(1):23–30

Malassis L (1983) Filières et systèmes agro-alimentaires. Économies et sociétés 17:911–921

Malassis L, Padilla M (1986) Économie agro-alimentaire, Tome 3 L'économie mondiale. Cujas, Paris, p 449

Marsh K, Bugusu B (2007) Food packaging – roles, materials, and environmental issues. J Food Sci 72(3):R39–R55

Masson S, Petiot R (2012) Attractivité territoriale, infrastructures logistiques et développement durable. Cahiers Scientifiques du Transport 61:63–90

McKinney ML (2006) Urbanization as a major cause of biotic homogenization. Biol Conserv 127(3):247–260

Mestdag I (2005) Disappearance of the traditional meal: temporal, social and spatial destructuration. Appetite 45(1):62–74

Monteiro CA, Cannon G (2012) The impact of transnational 'big food' companies on the South: a view from Brazil. PLoS Med 9(7):e1001252

Morana J, Gonzalez-Feliu J (2011) Le transport urbain vert de marchandises : leçons tirées de l'expérience de la ville de Padoue en Italie. Gestion 36(2):18–26

Odeye M, Bricas N (1985) À propos de l'évolution des styles alimentaires à Dakar. In: Bricas N, Courade G, Coussy J, Hugon P, Muchnik J (eds) Nourrir les villes en Afrique sub-saharienne, Coll. Villes et entreprises. L'Harmattan, Paris, pp 179–195

Ouattara I (2005) Les villes et les campagnes ivoiriennes : quels nouveaux liens? In: Coll JL, Guibbert JJ (eds) L'aménagement au défi de la décentralisation en Afrique de l'Ouest, Coll. Villes et Territoires. Presses Universitaires du Mirail, Toulouse, pp 149–162

Pimentel D, Pimentel MH (eds) (2007) Food, energy, and society, 3rd edn. CRC Press, Boca Raton, p 380

Pingali P (2007) Westernization of Asian diets and the transformation of food systems: implications for research and policy. Food Policy 32(3):281–298

Popkin BM (1999) Urbanization, lifestyle changes and the nutrition transition. World Dev 27(11):1905–1916

Porkka M, Kummu M, Siebert S, Varis O (2013) From food insufficiency towards trade dependency: a historical analysis of global food availability. PLoS One 8(12):e82714

Poulain JP (2002) Sociologies de l'alimentation : les mangeurs et l'espace social alimentaire. PUF, Paris

Poulain JP, Corbeau JP (2002) Penser l'alimentation. Entre imaginaire et rationalité. Privat, Toulouse

Rastoin JL, Ghersi G (2010) Le système alimentaire mondial : concepts et méthodes, analyses et dynamiques. Éditions Quæ, Paris

Requier-Desjardins D (1989) L'alimentation en Afrique : manger ce qu'on peut produire. L'exemple de la consommation alimentaire en Côte d'Ivoire. Karthala, Paris, 169 p

Ritzer G (2011) The McDonaldization of society. Pine Forge Press, Los Angeles

Rosa H (2010) Accélération : une critique sociale du temps. La Découverte, Paris, 474 p

Salomon CJ, Marchand B (eds) (2010) Antiurbain. Origines et conséquences de l'urbaphobie. PPUR, Lausanne, 329 p

Salomon CJ, Mathieu N (2014) Interroger une représentation collective : la ville mal-aimée. In: Martouzet D (ed) Ville aimable. PUFR, Tours, pp 125–154

Sankale M, Wone I, Morosov T, Moroso S, De Lauture H (1980) La place du « ceebu-jën » dans l'alimentation des populations suburbaines de Dakar. Présence Africaine, 1980/1(113):9–44. https://doi.org/10.3917/presa.113.0009

Satterthwaite D, McGranahan G, Tacoli C (2010) Urbanization and its implications for food and farming. Philos Trans R Soc B: Biol Sci 365(1554):2809–2820

Scardigli V (1983) La consommation, culture du quotidien. PUF, Paris, p 35

Schwartz B (2004) The paradox of choice: why less is more. Ecco, New York

Slovic P (1987) Perception of risk. Science 236(4799):280–285

Smith A, Watkiss P, Tweddle G, McKinnon A, Browne M, Hunt A, Treleven C, Nash C, Cross S (2005) The validity of food miles as an indicator of sustainable development – final report produced for Defra. UK Defra, Report ED50254, p 117

Sobal J, Khan LK, Bisogni C (1998) A conceptual model of the food and nutrition system. Soc Sci Med 47(7):853–863

Soler LG, Requillart V, Trystram G (2011) Organisation industrielle et durabilité. In: Esnouf C, Russel M, Bricas N (eds) Pour une alimentation durable. Réflexion stratégique DuALIne. Quæ, Paris, pp 109–142

Stuckler D, Nestle M (2012) Big food, food systems, and global health. PLoS Med 9(6):e1001242

Sudrie O (1985) Dépendance alimentaire et urbanisation, une relation controversée. Revue Tiers Monde 26(104):861–878

Tacoli C (1998) Rural-urban interactions: a guide to the literature. Environ Urban 10:147–166

Tibère L (2009) L'alimentation dans le « vivre ensemble » multiculturel : l'exemple de La Réunion. L'Harmattan, Paris

Touré A (1981) La civilisation quotidienne en Côte d'Ivoire. Procès d'occidentalisation. Karthala, Paris, 279 p

Tukker A, Jansen B (2006) Environmental impacts of products: a detailed review of studies. J Ind Ecol 10(3):159–182

Van der Spek RJ (2008) Feeding Hellenistic Seleucia on the Tigris and Babylon. In: Alston R, Van Niif OM (eds) Feeding the ancient city. Peeters, Leuven, pp 33–45

Viollet PL (2004) L'hydraulique dans les civilisations anciennes : 5 000 ans d'histoire. Presses des Ponts, Paris

Weatherspoon DD, Reardon T (2003) The rise of supermarkets in Africa: implications for agrifood systems and the rural poor. Dev Pol Rev 21(3):333–355

Webster D (1989) Coca-colonisation and national cultures. Over Here 9(2):64–75

Wilkinson J (2008) The food processing industry, globalization and developing countries. In: McCullough EB, Pingali PL, Stamoulis KG (eds) The transformation of agri-food systems: globalization, supply chains and smallholder farmers. FAO, Agricultural Development Economics Div, Rome, pp 87–108

Winarno FG, Allain A (1991) Street foods in developing countries: lessons from Asia. Food Nutr Agric 1:11–18

History of Urban Food Policy in Europe, from the Ancient City to the Industrial City

Benoit Daviron, Coline Perrin, and Christophe-Toussaint Soulard with François Menant (Contributor)

"What distinguishes and indeed contrasts the nation system and the city system is their structural organization. The city state avoided carrying the heavy burden of the so-called primary sector: Venice, Genoa and Amsterdam consumed grain, oil, salt, meat, etc., acquired through foreign trading; they received from the outside world the wood, raw materials and even a number of the manufactured products they used. It was of little concern to them by whom, or by what methods, archaic or modern, these goods were produced: they were content simply to accept them at the end of the trade circuit, wherever agents or local merchants had stocked them on their behalf. Most if not all of the primary sector on which such cities' subsistence and even their luxuries depended lay well outside their walls; and laboured on their behalf without their needing to be concerned in the economic and social problems of production. In all likelihood, the cities were but dimly aware of the advantages this brought and rather more conscious of the drawbacks: obsessed with their dependence on foreign countries (although in reality such was the power of money that this was reduced to almost nothing), all leading cities desperately tried to expand their territory and to develop their agriculture and industry. What kind of agriculture and industry though? The richest and most profitable of course. Since Florence had to import food anyway, why not import Sicilian grain, and grow vines and olives on the hills of Tuscany?" (Braudel 1984a: 295)

In this chapter we focus on urban policies in the field of food and agriculture, in Europe, from antiquity to World War II, seeking to relate the current debates discussed in this book to long-term developments. In so doing, our primary objective is to show how very diverse are the actions taken under urban food policies and thus to enlarge the range of possibilities considered in the current debates. Our second

B. Daviron (✉)
UMR MOÏSA, Cirad, MONTPELLIER CEDEX 5, France
e-mail: benoit.daviron@cirad.fr

C. Perrin · C.-T. Soulard
INRA, UMR Innovation, Montpellier, France

© The Author(s) 2019
C. Brand et al. (eds.), *Designing Urban Food Policies*, Urban Agriculture,
https://doi.org/10.1007/978-3-030-13958-2_2

27

objective is to reappraise cities' role in comparison with that of States, which have gradually acquired the status of major players in the area of food.

This historical approach reflects a certain diversity but makes no claim to completeness. European cities' relation to agriculture and food takes myriad shapes. One must in any case distinguish the two tendencies that have shaped urban Europe over the long term: 'central place' cities and 'network' cities (Hohenberg and Lees 1995: 4–7).

Under the central place model, the city provides services to (and administers) its surrounding area. That model produces a hierarchical urban structure, with many small towns, a diminishing number of larger centres, and a single capital—an organization that is very stable over time and may be implemented top down or bottom up.

Under the network model, the city participates in commercial, informational and political exchanges that go well beyond the region or the country and depend on the existence of (tangible or intangible) communications facilities. Distance is of little consequence. The network model promotes specialization and division of labour between cities and is exemplified by the emergence of industrial and mining cities. Such a hierarchy is unstable because of competition between cities of the network and the specializations they develop.

Both models are found in Europe. During the middle ages, the network model spread gradually from northern Italy, to Germany, to Flanders. Fernand Braudel writes: "[…] The destinies of these very special cities were linked not only to the progress of the surrounding countryside but to international trade. They were indeed to free themselves from rural societies and outdated political ties" (Braudel 1984b: 511).

Conversely, on either side of that corridor, to the northeast and southwest, the 'central place' model held sway. Both models were subject to imperial or national interference, whether through tax levies or predatory actions. The 'prince' would make his coercive powers available to a city, either directly, by commandeering products (episodically, or permanently, as tribute), or indirectly, by levying taxes.

Given that very diverse backdrop, we shall trace the history of urban food policy under three headings: first, the stages of Europe's urbanization. Second, cities' supply policies from antiquity to the modern era. And third and last, cities' public health policies at the dawn of the Industrial Revolution. Under the last two headings we shall be noting the progressive decline of urban policies because of rising State involvement. To conclude, we shall look at how cities could once again take food supply in hand.

The Four Stages of Urbanization in Europe

Four stages can be discerned in the history of European urbanization (Bairoch 1985; Hohenberg and Lees 1995; Mumford 1989). The first corresponds to the Greco-Roman civilization. It is marked by issues of supply for the imperial cities of the Mediterranean. The second stage begins in the middle ages, the golden age of cities as political entities with their own ambitious food policies. The Black Death (1347–1352) marks the start of the third stage, which comprises over three centuries of slow, sporadic urban growth (Hohenberg and Lees 1995: 6), and is marked by the emergence of nation States and a decline in urban food policies. The fourth stage begins in the mid-eighteenth century, with industrialization and the development of public health policies. Following our presentation of these four stages in the history of urbanization, we shall be looking at food policies.

Greco-Roman Urban Civilization

Ancient Greece was urban. At its height, c. 500 BCE, Athens seemingly had a population of some 100,000, while each of a number of other Greek cities had around 40,000 inhabitants. If we include towns with more than 5000 inhabitants, the urbanization rate may have been around 15–25%. The same is true, on a larger scale, of the Roman Empire. The population of the city of Rome alone appears to have been in excess of one million around the second century CE.

The end of antiquity is conventionally dated to the sack of Rome in 410. In the ensuing centuries, Roman civilization declined owing to the combined impact of invasions. Between 200 and 600, the population of Europe fell from 40–55 million to 20–35 million and became markedly less urban. The change in Rome's population is an extreme example of that process, as it fell to 50,000 by 700 and only some 30,000 by 1000 CE, despite the presence of the papacy. Other towns disappeared completely.

The Golden Age of Mediaeval Cities

Beginning around the year 800, the population of Europe, including the cities, began to grow again. Until the year 1000, the effects of urbanization were felt mainly in the regions of southern Europe that were part of the Arab world (Spain, Sicily), which at that time accounted for 50% of the population of European cities of more than 20,000. In that year, Cordoba and Palermo were the largest cities in Europe after Constantinople, the largest cities in Europe, with respectively 450,000 and 75,000 inhabitants (Chandler and Fox 2013).

In the rest of Europe, the urban renaissance began in the tenth century: "[…] thanks to the emergence of two foci, one in southern and one in northern Europe: Venice and southern Italy, on the one hand, and the Flemish coast on the other hand" (Pirenne 1927: 75).

Between 1000 and 1300, the population of Venice grew from 45,000 to 110,000 inhabitants, Genoa's from 15,000 to 100,000, Milan's from 30,000 to 100,000, and that of Florence from 13,000 to 60,000; while in Flanders, Ghent, Bruges and Ypres totalled 220,000 inhabitants in the middle of the fourteenth century. In the latter case, the growth was linked to the wool industry.

The urban renaissance then spread continent-wide, favoured by the resurgence of trade within Europe (especially between North and South) and with the Orient. The peak Roman-era population was quickly outstripped. Between 1000 and 1300, the total population, and that of cities, doubled, as the number of cities with more than 20,000 inhabitants rose from 35–45 to 100–110. Most of today's large cities emerged at this time.

Population growth came to a halt, however, at the beginning of the fourteenth century. A string of poor harvests caused famines, and then the Black Death struck Europe in 1347 and, over the next 5 years, killed a third of the population. Europe paid a heavy price for its resumption of trade: the first plague outbreak was in Caffa, a Genoese outpost north of the Black Sea (McNeill 1976). The plague then spread rapidly along trade routes, reaching Scandinavia 2 years later.

Slow Urban Growth in the Modern Era with the Rise of Nation-States

Population growth resumed in Europe after 1400. By 1500 the population had returned to pre-Black Death levels, but growth was slow and hesitant: over two centuries, it increased from 76 to 102 million. City development was in step with that increase, with no significant increase in the rate of urbanization. Large cities, indeed, save only London, had slower growth from the sixteenth to the seventeenth century. Charles Tilly (1990) believes that this was the consequence of a shift of industrial activity to small towns and countrysides, as merchants and entrepreneurs sought to escape the rules of town-based guilds. As a result, though new settlements did spring up, they were unchartered.

This new phase of urbanization was shaped, in particular, by two processes:

- First, the geographical centre of gravity of long-distance trade moved from the Mediterranean to the Atlantic, benefiting the United Provinces, where by 1700 urbanization exceeded 40%. The population of London grew tenfold between 1500 and 1700, from 50,000 to 550,000 inhabitants. The same was true, to a lesser extent, of other cities: Seville (up to 1600) and Lisbon, the ports of entry for products shipped from America to Spain; and in France, Nantes and Bordeaux.
- As regards the distribution of power, the period is characterized by the raise of territorial States, their growing grip on urban life, and the decline of cities' autonomy. Mercantilism (in Adam Smith's parlance) prevailed: sovereigns promoted the establishment of national markets by limiting the freedom of long-distance trade and by eliminating obstacles to internal circulation (e.g. by removing customs offices and tollbooths).

The sixteenth century marked a turning point. Speaking of the cities of the Mediterranean, Fernand Braudel wrote:

"[...] in the long term, they were healthy, since they were still growing: at any rate they overcame crises and difficulties; however, all towns without exception saw their liberties bring whittled away by the extension of the territorial states, which were expanding even more rapidly than the towns, surrounding them, subjugating them, or even chasing them from acquired positions. A new political and economic age was beginning." (Braudel 1995: 326).

In a discussion of the lengthy urban policy crisis in the sixteenth century, Braudel also wonders: "What disappeared in the course of this prolonged crisis? The mediaeval town, the city state, mistress of her own fate, set in the centre of her surrounding gardens, orchards, vineyards, wheatfields, and nearby coasts and roads." (Braudel 1995: 345).

In the seventeenth century, at a time when economies and populations were shrinking, the capitals of the then emerging territorial States administered by sovereigns, such as Paris, Madrid or St. Petersburg, became distinctly more populous than other cities. London, which was at once the seat of the Atlantic merchants and a royal city, became the second largest in Europe after Constantinople.

The Industrial Revolution Spurs Strong Urban Growth

As of the 1700–1750 period, a new phase of growth began with the rise of the cities of the Industrial Revolution: Europe's urban population grew sixfold during the nineteenth century. In both London and Paris the population grew to more than a million. In 1900, Europe's average urbanization rate was nearly 40%. "The main elements in the new urban complex were: the factory, the railroad, and the slum" (Mumford 1989: 657).

Mining and steel-making areas became important places of settlement: the German Ruhr, the Black Country of England, Borinage in Belgium, and in France the Pas-de-Calais and the Loire coalfields. It was there that the industrial city or 'coketown' as Mumford, after Dickens, calls it (Mumford 1961: 642 et seq.), emerged.

Industrial activity also developed in the older cities. Thanks to coal and steam power, with thermal (metallurgy, chemistry, etc.) and mechanical energy (spinning, weaving, machining, etc.), industry shed its dependence on wood, wind and watercourses. Steam engines also made it easier to supply cities with food, fetched from farther and farther afield. Long-distance supply had previously been much more complex, and had since antiquity needed rulers' direct involvement to ensure its safety.

Securing Food Supply: Rise and Fall of Urban Policies

"From antiquity to the modern day, feeding city dwellers to ensure political tranquillity and social stability has been a constant concern for rulers, and in discharging that function they forged strong bonds between the peoples and their sovereigns." (Marin and Virloulet 2003).

Special Features of the Mediterranean Food Supply

The Mediterranean is one part of the world where, since ancient times, urban governments have taken a direct role in long-distance supply, in particular of cereals. The overview by Marin and Virloulet (2003), *Nourrir les cités en Méditerranée – Antiquités – Temps modernes*, shows that supplies of wheat were a constant concern but that urban governments were never involved in the supply of all commodities: the private trading system still dominated trade. The reasons for cities to take an active public role varied by city, by State and by era: military strategy could be involved, or prestige, food quality, or control of famine and conflicts.

In the Mediterranean, the economic structures that governed city food supply underwent little change between antiquity and the modern period: poor or chancy agricultural yields, with unchanging transport and storage arrangements. The Mediterranean was still the geographical framework of the food trade. Cities set up wheat boards or 'offices of abundance' (public granaries)—institutions specifically intended to see that city dwellers received the grain they needed and so to forestall food shortages and starvation. But, while grain supply to the cities of the North was mainly organized by private agents (except for sporadic price-fixing), in all Mediterranean cities there was at once a private supply and a public *annona* (Grantham and Sarget 1997).

Thus, the history of the Mediterranean affords us valuable information on cities' strategies and ways of influencing grain supply markets. The foundational *annona* system of ancient Rome is the guiding thread for this first historical period. Subsequently we shall see how the *annona* system evolved in modern times.

The Ancient Annona System in the Mediterranean

The principle of the ancient *annona* is as follows: magistrates have specific responsibility for monitoring the city's markets and must, should difficulties arise, buy foodstuffs and distribute them at cut prices or free of charge. Rome innovated by making these distributions, at first exceptional, a regular occurrence, transforming the *annona* into a civic institution, which would later be copied by Constantinople. It was a prerogative of the capitals, as the system existed only briefly in other cities.

Wheat redistribution by the *annona* was made possible by in-kind levies on certain provinces, such as Egypt and Africa, and the income of the Imperial domains. Ancient Rome drew its supplies chiefly from Egypt (33%), North Africa (10%), Sicily, Sardinia and the rest of Italy. The African share subsequently grew, as Constantinople developed and consumed a growing proportion of Egypt's production. The Black Sea region and Egypt were the sources of supply for Athens and Constantinople. The major Mediterranean ports were Pozzuoli, Alexandria, Narbonne, Cadiz, and Carthage.

The *praefecti annonae*, the Imperial officials who supervised the grain supply, had other duties as well: with a role to play in port infrastructure, incentives for carriers and merchants, and management of a special fund for merchandise purchases. The office of the *annona* also regulated foreign trade in grain, banning exports to forestall shortages or, when harvests were abundant, encouraging them instead, to head off a collapse of domestic prices. Periods of scarcity were marked by the sending of senatorial grain commissions, which would make emergency purchases of foodstuffs that were then distributed in the city.

A key role was played also by the great landowners (senators, knights) in possession of extensive agricultural estates, not just in Italy but also in the provinces of Sicily, Egypt and Africa. Often they owned warehouses, which they would let for a tidy profit.

The *annona* system contributed to ancient cities' strength, being founded on the links between prominent merchants and the ruling classes.

The Annona System in Modern Times

In modern times, the Mediterranean ceased to operate in a vacuum. The supply area had expanded. In the eighteenth century, Marseille became the largest Mediterranean port of trade for grain.

The modern *annona* differed from the ancient one (Marin and Ventura 2004): its role was no longer to regularly distribute free grain to part of the city's population. The modern equivalent was a marketing board, acting to regulate market prices: buying and storing grain in good years and, in bad ones, putting it back on the market at low prices, to drive prices down. It could also act to control bread-making, for example by regulating the number of ovens allowed for public sale of bread. Thus, organizations of this kind became graduated response systems that did not control the whole urban food supply.

They were caught between conflicting priorities—social peace and profit—and afforded many examples of collusion between officials and food supply professionals, constituting licit or illicit interest groups (Martinat 1999). These policies were also often based on a power balance between a city and a territory over which it sought to wield a monopoly, or at least to ensure that it had the pre-emptive right to buy agricultural commodities. In practice, however, a balance was created between consumer protection and the need to keep both producers and traders solvent.

From antiquity to the modern era, the *annona* system helped feed large urban populations in a world still dominated by rural life. That kind of balance was sustained for a long time in some cities (Constantine, Rome, Madrid), less so in others (Naples), but storage-based regulation of market prices put a great strain on municipal finances. For that reason, with the increasing urban population as one factor and widespread price inflation as another, a breakdown occurred in the second half of the eighteenth century as the old *annona* systems failed. Thenceforth those systems could no longer satisfactorily deal with cities' food supply, which States would now be called upon to regulate (Revel 1975).

The Golden Age of Urban Supply Policies: The Middle Ages in Europe

The mediaeval period was a golden age for urban policies, particularly food supply policies. For cities in the Middle Ages, facing the spectre of famine, food supply was a major challenge (Box 1). "[T]he regulation of food markets within the city was a major area of activity. It was very complex in detail and the authorities continuously adjusted rules to deal with practice and to meet the requirement of changing markets." (Keene 1998: 4).

Box 1: Famine, Markets and Public Policy in European Cities from the End of the Middle Ages

François Menant

Urban societies at risk of famine
Beginning in the twelfth century, but especially the 13th, Europe was undergoing rapid population growth, which was only finally halted by the great plague of 1348. Paris and a few Italian cities had more than 100,000 inhabitants; Catalonia and the Valencia region, Flanders, the Rhine Valley and a few other regions had dense urban networks. Much of their inhabitants' livelihood came from the salaries paid by industry, which at the time was achieving growth by exporting its products, including textiles, to European and Mediterranean markets. The people's staple food was bread: most of them would buy wheat in the market, have it ground at the mill, knead the dough themselves, and take it to the bakery to be baked.

Beginning in the 1270s, but especially after the 'Great Famine' that ravaged Flanders and England in 1314–1317, recurring food crises hit these poor workers, just like the poor rural masses. The deteriorating climate Europe was experiencing at the time certainly had something to do with this, even if it is not a complete explanation: poor harvests now happened every 4–5 years, causing more serious shortages. In Florence, there was an episode of inflation

---

(continued)

Box 1: (continued)

on average every 6 years between 1309 and 1375, with the price of wheat rising each time: its average decadal price, referred to an index of 100 in 1271–1286, reached 481 in 1339–1353.

Food shortages came to a head just before the plague of 1347 that affected all of Europe. Nor did they end thereafter, as might have been suggested by Malthus's theories. That goes to show that they were not simply an effect of overpopulation, as the population had in fact dropped sharply. Post-plague conditions were nevertheless a little better, as workers' wages had risen to a point where they could feed their families except in times of extreme shortage.

Mechanism of food crises

Europe now had a market economy in which shortages were reflected in rising and indeed soaring prices, known by the Latin word *carestia*. Poor harvests were only the trigger, or indeed pretext, for the inflation (Sen 1981). The decisive inflationary mechanism went as follows: when a crop failure was expected, at winter's end or even earlier sometimes, brokers—merchants and wealthy landowners—banking on rising prices, would keep their stocks off the market. Some grain would be released to market, but because quantities were limited, the prices stayed high.

The working poor, who had not been able to put anything by, had no option but to continue to buy when prices rose, and ended up spending much of their money, then all of it, on wheat. As a result, demand for manufactured goods, and in general everything but staple foods, temporarily plummeted: the crisis hit the entire economy.

Public food shortage policies

Beginning in the last third of the thirteenth century, in most cities and states, institutions and tools were put in place to lessen the seriousness of such crises. Supply from distant sources was now largely the responsibility of States and, especially, municipal authorities, who had the wherewithal to prevent and combat shortages: money, shipping and storage capacity, as well as intelligence on foreign production sites. It should be added that in the cities, the ruling circles were strongly motivated, on ideological and political grounds, to take action against scarcity: thus, Italian municipalities acted for the 'common good' or simply out of fear of the people's anger.

The great cities of the 14th and 15th centuries, especially those with easy access to the sea, were largely supplied by imports: Sicilian wheat fed Florence and the other cities of Tuscany; Venice was supplied by Puglia; Genoa from the far end of the Black Sea; and the cities of northern Germany from the eastern shores of the Baltic. In times of scarcity, imports played a bigger role, because it was very rare for the whole of Europe at once to be

(continued)

Box 1: (continued)

affected by a poor harvest of all cereals. The landing of a cargo, or even just the announcement of its arrival, was enough to ease market prices. As soon as they heard the news, those who had been holding back their stores hastened to put them on sale, to take advantage of the last days of high prices; which immediately drove those prices down.

Learning the lessons of the major food shortages of the late thirteenth century, urban authorities bought wheat as soon as concerns were raised about the harvest. At times, too, they would build up stores in advance, without waiting for warnings of scarcity, and sell them at moderate prices to those who needed them. However, the municipality's wheat would not be sold at too much of a discount to the market price, so as not to undercut commercial supply. Alternatively, the municipality could act to distribute wheat to bakers, or even produce bread itself, which was then sold at affordable prices. Charitable institutions (hospitals), which had extensive landholdings and monetary resources, also played a significant role in preventing mortality; in that sense they took over from the monasteries, whose histories reveal the charitable role they had played in times of famine since the early Middle Ages.

Another part of the supply management measures was coercive: producers and middlemen were forced to put their reserves on the market (which meant stock checks and searches) and forbidden to export, while prices could not be raised beyond a set ceiling; but the ceiling price led to wheat becoming unavailable on the market, and urban authorities soon learned to wield the double-edged weapon of price controls with caution.

Supply, then, especially in times of food shortage, was a capital issue for the cities of the late Middle Ages: a demographic and public health issue of course, but also a political and ideological one.

Mediaeval cities' food policy must be placed in the more general context of the five objectives pursued in their economic policy (Heckscher 2013: 128):

1. To guarantee the city abundant supply by preventing its supply from being diverted to another destination.
2. To prevent the development of rural activities that would compete with the city's. (These two objectives were sometimes pursued jointly, for example in Germany, where cities delimited an area on their outskirts that was reserved for agricultural production and in which manufacturing was forbidden.)
3. To ensure that local merchants participated in trade and to forbid any direct relationship between *outside* merchants and the rural population.
4. To do everything possible to ensure that trade took place within the city, for example by requiring unloading to be done in the river cities.
5. To guarantee everyone a livelihood in keeping with their social status, for which purpose it was vital to limit competition (role of guilds).

In the food industry, the first objective is what defines and distinguishes mediaeval cities' policies:

"[...] The dominating feeling throughout the Middle Ages, mostly in towns, which were almost the only repositories of medieval economic policy, was the one natural to consumers; they wanted to hamper or prevent exports but favoured imports; their tendency was a 'love of goods'; their policy may be called one of provision." (Heckscher 1936: 49).

Henri Pirenne (1917) outlined the legislation of the Netherlands' mediaeval cities. With the goal of "providing the townspeople with cheap, abundant foodstuffs", intermediaries were considered speculators. "It followed that the country seller and the urban buyer must be put in direct contact so as to prevent the market from being cornered by a group of speculators" (Pirenne, 1917:100). He cites the example of Liège, where, in 1317, the *Lettre des vénaux*:

"[...] forbids 'regraters' [middlemen] to buy poultry, cheeses or venison within a two-league radius of the city. All such foodstuffs must be brought to the market, and wholesale merchants shall be able to acquire any surplus only after the burgesses [townspeople] have been supplied therefrom. Butchers are forbidden to keep meat in cellars and bakers to buy more grain than they need 'for their own baking'." (Pirenne 1917: 100).

All possible measures are taken to keep food prices down: "Not only is a maximum price set, but it is absolutely forbidden to sell outside the market, i.e. anywhere except in public and under the supervision of the burgesses and officials of the municipality" (Pirenne 1917: 100).

The ban on traders' buying in the vicinity of the town and the requirement that trade take place in a public market were found in many towns. The purpose was both to eliminate traders' profit margins and to ensure that local products would not be sent elsewhere. In Paris, traders were forbidden to buy within a radius of 10 leagues in the case of wheat, 20 leagues for wine (Aymard 1983: 1404). In Geneva, as late as the seventeenth century, the city administration sought to make fairs and markets the only places where trade with farmers was permitted. A political decree or police order set a radius in which production was reserved for urban supply: producers had to sell in town, while buyers were not allowed to buy directly from producers within four leagues (18 km) of the city (Niwa 2015). In Venice, this policy of exclusive sale was taken to another level. In 1234, a treaty required the merchants of Ragusa to sell wheat and salt only in Venice. Another treaty, from 1236, forbade all trade in the northern Adriatic except the transport of food to Venice (Hibbert 1963: 174).

As another way of securing the food supply, taxes were often low or nonexistent on inbound goods, while those leaving were subject to high taxes, as in the German towns or in Florence. Such taxes were often price-adjusted, being imposed only when prices rose above a certain level (Hibbert 1963: 176).

Similarly, towns often had bylaws making their own people priority buyers or indeed the only ones entitled to buy for a set time after the products' arrival.

Finally, towns sought seek to avoid any takeover of supply by a monopoly or commercial cartel. In 1268, in Douai, there was a legal limit on the daily quantity (e.g. of wheat) a person could buy (Hibbert, *op. cit.*). Trade and processing of food products were often the last and only business sectors not subject to a guild monopoly.

Trade circuits could also be directly controlled. In 1555, in Valencia, Spain, there was an elected magistrate, the *clavarie del vituallement*, assisted by deputies who dealt with wheat and meat (*administradors dels forments* and *administradors del carns*). The municipality helped supply grain to the city. This took the form of wheat (*forment asegurat*) that it bought in bulk and sold at a fixed price, to keep the market price low (Lapeyre 1969: 134] cited in Tilly 1974: 438).

All major Italian cities also had a wheat office, and sometimes and oil office too: *Abbondansa* in Florence and Siena, *Officio delle Biave* in Venice. Maurice Aymard notes that all of them were in agreement that: "[…] the city's supply could not be left to private enterprise alone, for entrepreneurs had neither the means nor no doubt the concern to ensure the abundance and regularity of supply deemed considered necessary" (Aymard 1966: 72).

Because of its size and power, Venice exemplified in the extreme this commitment to the flow of foodstuffs. All grain entering the city was monitored. Locally produced grain could be sold only in Venice, and Venetian farm owners were required to send in their harvest by summer's end. Grain imported from farther afield was subject to long-term purchase contracts with merchants that included loans. A system of import premiums was provided for crisis situations (up to 40% of the domestic price). Purchases from the most distant growing regions (Burgundy, Bavaria, Poland) were made directly by the authorities. Importers were allowed to retain only a limited share for their own consumption. All the rest was put on sale in the two public shops that supplied flour to individuals to make their own bread. The bulk of the grain, however, was set aside for bakers, whose numbers were limited and who must comply with strict rules on the quality, weight and price of bread. The retail price remained fixed but the weight would vary depending on wholesale prices, which the authorities sought to control. Wheat could under exceptional circumstances be sold below market price.

At this period the cities were clearly seeking to keep the countryside in a subordinate role, as a source of biomass but also of labour, even though their relations were commercial in nature. The city often had a conflictual relationship with warlords and other neighbouring feudal powers, playing a middleman role and serving as a refuge for farmers seeking to evade their local lords' suzerainty. The cities had a high mortality rate and a structural population deficit that made them dependent on rural immigration.

Hence, urban wheat stores could be used to supply the countryside: in Geneva, as of the late seventeenth century, farmers received assistance in the form of seed wheat or bread whenever a subsistence crisis arose (Niwa 2015). In providing that assistance, the Wheat Board did not merely seek supplies from the countryside, but actively contributed to the development of agriculture.

Lastly, the existence of a nearby town provides an outlet for rural production. That, according to Max Weber, was what made the difference between the agrarian structures of Germany to the west and east of the Elbe. To the east, the lack of towns meant that only lords who could export grain over long distances (particularly to

Amsterdam) had access to the market. To the west, in contrast, a dense urban network allowed farmers to get rich and gradually achieve emancipation (Weber 1991).

Rise of the States and Decline of Urban Food Policy

In the modern era, the construction of the territorial States upset the urban food policies described above. States intervened more and more directly in food production and distribution beginning in the sixteenth century and up until the great wave of liberalization of the second half of the 19th. Unification of the national market was a goal of the territorial States, but one that went against all the measures the cities had adopted to secure their people's food supply. It was nevertheless eventually achieved. In France, unification was not complete before Colbert, despite the enactment of successive decrees by the predecessors of Louis XIV.

Again, under mercantilism, the territorial States adopted strategies of autarky (Clément 1999: 38) that jeopardized urban supply policies based on long-distance trade. Moreover, the States, having responsibility for the government of cities and countryside alike, pursued policies that skewed more toward producers' interests, while consumers' interests tended to be overshadowed. Thus, in England, the first national measures taken at the end of the seventeenth century sought to promote grain exports, an orientation that was diametrically opposed to what mediaeval cities wanted.

This rise of the States continued well beyond the eighteenth century. It is true that, in the nineteenth century, Europe experienced a "free trade interlude" (Bairoch 1993: 39), which began in 1846 with the abolition of the United Kingdom's Corn Laws, but it did not last long. As early as 1870, continental Europe adopted protectionist policies and the two world wars as well as the crisis of the 1930s reinforced State intervention: creation of domestic marketing boards, proliferation of border controls on foreign trade, food aid distribution, etc. This state of affairs was perpetuated by the creation of the Common Market, then the EU's Common Agricultural Policy, which introduced a mechanism for protection of agricultural prices in the domestic market. States only withdrew from the management of food markets with the enactment of the European reforms of the 1990s and the World Trade Organization agreements.

As of the nineteenth century, another major area of intervention for urban policies emerged with the hygienist movement and the cities' separation from the organic world.

The Cities' Separation from the Organic World

"I have argued that the sanitary idea and its enthusiastic adoption by many in the public health movement were responsible for two major changes in the mid and later nineteenth century. First, there was a materialization in physical

infrastructure of the idea that waste products and their smells had to be removed before they could cause disease. A range of technologies, from sewers to waste destructors, were employed to achieve this purpose. Second, food-producing animals and animal by-product industries became unwelcome in many cities, with the ultimate aim of establishing nuisance-free, and therefore cleansed, environments. Together, these amounted to a greater conceptual and physical separation of the urban realm than had ever been experienced before." (Atkins 2012: 77).

In the nineteenth century, rapid urbanization caused health problems, as epidemics arose (McNeill 1976). Transportation development facilitated the spread of germs. Industrial activity increased pollution of the air, water, and even soil (Mumford 1989). Such pollution was the result of new technologies and the fact that cities now hosted insanitary premises that had previously been spread across the countryside. Finally, organic waste from food and the presence of animals in the city (particularly draught horses) was a potential source of contamination and proliferation of germs. As a result, excess urban mortality increased during the first phase of industrialization. In 1840, a man born and living in the country (County of Surrey) had a life expectancy of 44 years, compared to 35 for a Londoner and 24 for a Mancunian. Such large discrepancies were related, in particular, to infant mortality in cities, where infectious diseases (typhoid, tuberculosis, pneumonia, cholera) and food- and waterborne illness were frequent.

By the end of the eighteenth century, that situation had given rise to discussions that would soon engender hygienist policies. The 'progress' of science and technology was paralleled in medicine by the promotion of public health. The goal was prevention rather than cure, a cleansing of the environment (Jorland 2013: 24). This 'century of hygiene' spanned the last decades of the eighteenth century to the first of the 20th (Frioux et al. 2011).

We shall illustrate that shift by taking up the following topics successively: the regulation of food wholesomeness, the issue of animals' place in the city, and organic waste management.

Regulation of Food Wholesomeness

Urban authorities (municipalities) were important promoters of hygienist policies, but gradually the national public authority (State) asserted itself. Healthful food was at first a concern of municipalities, which at that time had lost all their prerogatives regarding market price regulation. But, by the turn of the twentieth century, in this area too the State had evinced the cities, which became mere local agents of the State, which, through legislation, assigned them a great part of their duties and obligations.

The wholesomeness or nutritional quality of food is not always easy to assess. The presence of pathogens or adjuvants is detectable only by complex techniques. In 1909, Sheridan Davis, the officer in charge of Manchester's milk supply, said that:

"The inhabitant of large towns has generally no access to the sources of his food; he is seldom able to satisfy himself by personal observation of the genuineness and soundness of the articles offered to him for consumption, and has to depend upon the knowledge, skill and vigilance of persons appointed for the purpose of protecting him against the dangers associated with insufficient or unsound food supplies" (Delépine, quoted in Otter 2006: 526).

In the nineteenth century, however, the development of chemistry and biology considerably expanded the range of analytical methods and of risks to be assessed. Thus, typhoid contagion through milk had been demonstrated by the 1870s, while the transmission of bovine tuberculosis to humans, in milk or by meat eating, was the subject of very serious concern. That was the backdrop for the establishment in many European cities of municipal laboratories to monitor product quality (Frioux et al. 2011) (Box 2).

Box 2: Development of Analytical Laboratories in France

Benoit Daviron

In France, a 1790 ordinance gives as one of the municipal police's powers: "inspection of fair dealing in measuring commodities sold by weight, length or volume and of the safety of foodstuffs offered for public sale" (Paquy 2004: 45).

At the beginning of the nineteenth century, at the Restoration, 'wholesomeness councils' were created in various cities (Lyon 1822, 1825 Marseille, Lille 1828…). They were made mandatory by the central government in 1848. Then, in the late 1870s, municipal hygiene offices proliferated. In particular, they looked into the wholesomeness of food and drink.

The history of the Grenoble laboratory studied by Lucie Paquy is a case in point (Paquy 2004). In Grenoble, the idea of creating an analytical laboratory to combat food fraud and adulteration was first floated by City Council in 1881. Its implementation drew on several major cities' experience with such laboratories: Brussels (1856), Antwerp, Paris, Lyon, Saint-Étienne, Brest, Le Havre, Montpellier, Le Mans and Marseille (Tanguy 2007). At first, the qualitative and quantitative analyses people requested were billed to them, to finance the laboratory and avoid to great a number of requests. In that way, Grenoble differed from Paris, Lyon, Toulouse or even Le Havre, where analysis was based on the existence of permanent inspection and sampling services.

(continued)

Box 2: (continued)

"In Lyon, for instance, inspection of foodstuffs, and more broadly of 'any item whose use may have health effects', was the responsibility of the laboratory's four expert inspectors, who had the police provide samples of suspect products. These samples were then sent to the lab for analysis and, if fraud was detected, the municipality referred the case to the Prosecutor's office. The effect was that Lyon was able to continuously monitor product quality" (Tanguy 2007: 50).

During the first few years, wines made up nearly 80% of the samples analysed. Among the other products were milk (skim or 'blue'), vegetable oils (often mixed), butter and liquors.

But the law of 1 August 1905 relating to "the suppression of fraud in the sale of goods and falsification of foodstuffs and agricultural products" redefined the division of powers between the State and the municipalities. It implemented a central fraud unit at the Department of Agriculture. Prefects were given the responsibility of organizing local collection and analysis services. Hence, they appointed collection officers who would send samples to the lab and, if fraud was suspected, refer the matter to the Prosecutor's office.

The communes were expected to play a role in stamping out fraud. The laboratories set up by the municipalities were, like the State labs, authorized to perform tests, provided they had been previously approved by the Ministry of Agriculture. The Grenoble laboratory received certification for the departments of Isère and Hautes-Alpes on 18 April 1908. In 1910, however, after some faulty analyses, the Ministry of Agriculture decided to temporarily suspend its certification and made reorganization recommendations to the municipality of Grenoble. The laboratory's operating procedure and staff recruitment process were increasingly set by Paris. Before the First World War, the municipal laboratory mainly survived on State subsidies and became the local authority for the repression of fraud.

Debates and regulations on animals' proper place in the city formed part of the same hygienist movement.

Regulation of Animals' Presence in the City

From Private Slaughterhouses to Industrial Municipal Abattoirs

At the beginning of the nineteenth century, the slaughter of animals was done at a a large number of private, cottage-scale slaughterhouses located in the city close to consumers. The public authorities progressively took over their activity. The French term 'abattoir' is said to have been coined to designate slaughterhouses commissioned by the authorities: the large-scale ones built from 1806 to supply Napoleon's troops (Otter 2005: 95). An 1838 ordinance classified slaughterhouses among "the most unhealthy, dangerous and noxious places", but the regulations relating thereto were the responsibility of municipalities, which set their own health policies. In the course of the nineteenth century, municipalities began building and managing their own abattoirs (Muller 2004: 107).

Séverin Muller reports on the process of creating a municipal abattoir at Saint-Maixent-l'École in the 1860s. The municipality's goal was to limit nuisances, such as odours and contamination. Strict construction rules were laid down to ensure hygiene. Slaughterhouse inspection was the responsibility of veterinarians employed by an 'animal health police' organized by the department, but in practice field inspections were carried out by municipal roadmen, police officers, or the establishment's security guards. In Saint-Maixent-l'École, because the municipality could not afford a full-time veterinarian, a 'veterinary artist' was hired to inspect the abattoir every two days and issue slaughter permits for animals certified healthy. Continuous monitoring of the slaughterhouse was done by a *garde champêtre*. As time went on, the slaughterhouse director too was tasked with examining cows raised within the commune to make sure they did not have diseases that could contaminate the milk.

In the United Kingdom, the same abattoir municipalization movement got under way later. In 1870, there were still 1500 private slaughterhouses in the city of London alone (Atkins 2012: 84). The slaughterhouses were most often unspecialized spaces that were occasionally used for animal slaughter. Being so numerous, and not purpose-built, they were virtually impossible to inspect. The Public Health Act of 1875 directed local authorities to create public abattoirs, and in 1890 a new law authorized them to shut down insanitary slaughterhouses.

Once Ubiquitous, Animals Are Expelled from the City

Before they were expelled, animals were everywhere in nineteenth-century industrial towns. "[I]t is possible to argue that animals were constitutive of a certain stage of the urban. They facilitated growth, they fuelled it, and they provided an essential continuing link with the parallel rural economy." (Atkins 2012: 35).

A number of factors contributed to animals' presence in the city. Most important was the perishability of fresh products, milk in particular, in the absence of refrigeration technology. From 9000 at the end of the eighteenth century, the number of cows in London had increased to 15,000 by the mid-19th; they were fed on the by-products of breweries and distilleries. For urban consumers, local supply also offered some guarantee against the frequent practice of watering down the milk.

The next greatest reason was urban growth, which required unprecedented numbers of carriage horses for local transport. In London, the number of horse-drawn cabs rose from 1265 at the beginning of the 1830s to 6800 in 1863 and 11,000 in 1888 (Thompson), 1976). Horse buses numbered 620 in 1839, whereas in 1902 there were 3696 (with 11 horses per bus, as they needed rest!). In Paris, the number of horses per 1000 inhabitants rose from around 23 to 35 between 1820 and 1880 (Barles 2012: 175).

Finally, the proliferation of animals in the city was encouraged by massive influxes of country dwellers, who brought their livestock with them and often kept up an agricultural activity while adapting to city life. In Manchester, Irish migrants were criticized for keeping pigs in town, even though that practice was actually common well before the great waves of immigration (Scola and Scola 1992: 39).

All of these factors contributed to the omnipresence of animals in the city, a situation that Peter Atkins called "urb-an-imal" or "animal-constituted urbanism" (Atkins 2012).

In a subsequent phase, however, after 1850, more and more measures were taken to ban animals from the city. Particularly complained of were the physical danger they posed to passers-by (Barles 2012: 182 et seq.), noise, odours, and especially the risk of contamination and disease. Thus, urban cowsheds where the cows lived tightly packed together and never got out (until they left for the slaughterhouse) were vilified as reservoirs of typhoid, scarlet fever, diphtheria and tuberculosis.

Atkins gives a chronology of the London by-laws adopted from 1844 on that little by little eliminated animals' place in the city (Atkins 2012: 28–33). The Metropolis Management Amendment Act of 1862, for instance, reaffirmed urban authorities' power to close unfit pigsties and prevent their keepers from opening new premises. It also introduced a compulsory licensing system for cowsheds. A fuller set of measures was created following the enactment of the Public Health Act of 1875.

Live animal markets were also outlawed. In London, Smithfield Market, which had been in operation since the tenth century, was moved in 1855. By the end of the nineteenth century, standards had been adopted for cowsheds setting out construction materials, the type of openings required, and a minimum space per animal. A monitoring system was established. In 1900, the city of Glasgow inspected 1220 cowsheds and 13,919 cows (Otter 2006: 525).

In London, the number of cows fell from 15,000 in the mid-nineteenth century to no more than 3000 on the eve of the First World War. London's self-sufficiency in milk fell sharply: from 80% in 1850 to 28% in 1880 and just 3% in 1910 (Atkins 2012: 41). Of swine in Manchester, Roger Scola writes:

"[…] While as late as 1866 the sanitary authorities of [Manchester] remarked on the persistence of 'a passion or infatuation amongst very many of the working classes for pig breeding and pig fattening', only ten years on they were able to report that they were meeting less resistance in their efforts to clear away the pigs, even from the poorest areas." (Scola and Scola 1992: 40).

Thus, by the turn of the twentieth century the 'Great Separation' of human residence and animal production was accomplished (Atkins 2012: 2). Soon only pets would remain in the city. Regulations were compounded by successive innovations in the conditions of transportation and product storage. Fresh produce could now be grown in places remote from consumers, and as the automobile gained ground, horses too were banned from the city.

The drive to rid the city of its organic waste dates from the same era.

Organic Waste Management

The invention of urban waste is recent (Barles 2005a). Only when the city reached a certain size and its links to the countryside and agriculture loosened, so that waste was no longer mainly organic, did waste management become a problem, requiring a public policy and municipal management.

Ridding the City of Its Waste

The best-known ancient sewers are probably those of ancient Rome, which are partly still functional. An open-air ditch (the *cloaca maxima*) was built by Tarquin the Elder in the sixth century BCE, at a time of rapid population growth, to drain stormwater as well as wastewater into the Tiber. The canal, probably inspired by Etruscan techniques, also helped drain standing water between the Palatine and Capitoline Hills, a lowland area that then became the site of the Circus Maximus. By the second century BCE, it had already become a real underground sewer. These major urban projects of ancient Rome were built by Imperial command, so the term 'municipal policy' was not yet apposite.

Most Western cities took over waste management as one of their prerogatives in the second half of the nineteenth century, following the Industrial Revolution, with the goal of cleaning up the city, to get rid of waste that was deemed a source of disease. Engineers were then called upon to come up with technical solutions whereby the city could efficiently clear away a daily increasing flow of waste.

In London, wastewater was at first discharged directly into the Thames with the rainwater, a recommendation of the Consolidated Commission of Sewers created in 1848 (Trench and Hillman 1984). Only in 1854 was the connection made between poor water quality and the recurrence of cholera epidemics. A new, high-capacity underground sewer system was then designed by the engineer Joseph Bazalgette. Parliament voted to finance the project after the 'great stink summer' of 1858. The work was entrusted to the Metropolitan Board of Works. Wastewater was collected by a gravity sewer and flowed into the Thames estuary, where it was raised into reservoirs by monumental pumping stations (driven by steam engines). The reservoirs emptied into the sea by tidal action. The new sewer system, personally inaugurated in 1865 by the Prince of Wales, made cholera epidemics a thing of the past and cleaned up the Thames. As it was a Parliamentary project, it was directly managed by the London County Council beginning in 1889.

In Paris, wastewater emptied into the Seine, the Bièvre or Ménilmontant Brook. As in London, following recurring cholera epidemics and pollution of the Seine, a modern underground sewer system was created during the great urban transformation undertaken by Haussmann as Prefect of the Seine from 1853 to 1870 (Reid 1993). For the first time, town planning was being done according to a comprehensive plan grounded in hygienic principles: the opening up of the Grands Boulevards cleansed and aired out the urban fabric. Underground conduits followed the line of the new streets. The sewer system was designed by Eugène Belgrand, a civil engineer polytechnician who in 1867 became Director of Paris's water and sewerage department. It is a combined gravity-fed sewer (collecting runoff and sewage). Progressively, buildings were obliged to connect to it: in 1852, an Imperial decree provided that: "[…] any new construction, in a street with sewers, must be configured so that rainwater and greywater feed into them" (Haussmann's memoirs, cited by Barles 2005b).

An 1894 law prohibited direct discharges into the Seine. The first sewage outfall was at Clichy, downstream from Paris, in settling ponds. The drainage channels were extended to Achères in 1895 (Reid 1993). In Paris and London both, it was hard to draw a line between State and municipal policies. In both cases the sewers were funded by the central government, a key role being played by a few expert engineers (Barles 2005b). Once the infrastructure was in place, it was managed by a municipal engineering department (the County in London, the Prefecture of the Seine in Paris).

As this hygienist period unfolded, a new vision of nature emerged: romantic, clean, sanitized. It advocated the physical separation of city and agriculture. And yet some urban waste continued to be recycled to feed the city.

Waste Recycling to Feed the City

Waste management is a specifically urban problem, one municipal authorities are faced with only when the city exceeds a certain size. Before it became a problem, waste was considered a resource.

Until the nineteenth century, self-regulation and recycling held sway: waste was organic and was recycled in agriculture as fertilizer, as animal feed (pigs, chickens) and sometimes in industry (dyes, papermaking pulp). Swine ran free in the city, feeding in the streets on domestic waste that residents were encouraged to leave for them (Clubbe 1992). In London, however, ordinances were adopted as early as 1357 prohibiting residents from throwing their garbage in the streets (Trench and Hillman, 1984).

The *jardins maraîchers* (market gardens) of Paris illustrate the dual role played by agriculture: as a food source and a waste disposal.

The French term *maraîcher* is derived from the *marais* (marshes) on the right bank of the Seine (today's Marais district) occupied from the twelfth century by farmers who drained the marshes to produce fruit and vegetables and sold them at the neighbouring market, Les Halles (Quellier 2015). Eventually, under the Ancien Régime, the term *marais* came to be understood as any market garden, no matter where it was or what kind of soil it grew on. Indeed, these areas of agricultural production were pushed out to the suburbs as the city expanded and the fortifications were rebuilt farther out, but links to the city remained close, for both marketing and fertilization.

Up until the eighteenth century, the contents of Paris's cesspools were stored and dried at the foot of the Buttes-Chaumont and then sold to farmers as fertilizer (Guerrand 1985). Up until the end of the nineteenth century, when farmers came to sell their products at the Halles market, they collected organic waste, including horse manure (Taylor-Leduc 2015). It was the same in London (Thick 1998). In Paris in the 1870s, some market gardeners had an actual contract with the municipality to pick up sludge at every street corner. These nightsoil collectors were granted a 3-year licence to ply their trade in a given neighbourhood by the adjudicators whose job it was to have the sludge and refuse of Paris removed. The sludge, a mixture of earth, water, manure, straw and domestic waste, was particularly rich in chemical elements (Barles 2005a).

Once the sewer system was in place, agriculture again had a role to play in purifying sewage before its discharge into the Seine. The engineer A. Mille drew on what Edwin Chadwick had done in London to develop the landfarming system. After a promising experiment near the Clichy outfall, which was deemed successful, the Paris water and sewerage department reached an agreement in 1869 with the commune of Gennevilliers for an expansion of the landfarming area, which went from 6 ha to 115 in 1874, 295 in 1876, 379 in 1878, and 422 in 1880 (Barles 2005b). This rapid expansion shows the success of an organization that was deemed: "[…] economically viable, agronomically effective, and hygienically satisfactory as it removes nuisances from the capital" (*ibid.*: 71).

Though physicians feared the sludge spread on farmland would cause disease, the strategy was immediately successful. Crop irrigation did in fact filter the water and protect the Seine. Barren gravel soils became fertile. The sleepy little town of Gennevilliers turned into a garden of Eden. Emperor Napoleon III visited the town and came back laden with vegetables (Steel 2008). By 1879, about 10% of the fruits and vegetables at the Halles market came from landfarming sites; often they were sold cheaper than the produce of conventional farmers because yields per hectare were greater (Taylor-Leduc 2015). So successful was the system that it was adopted Europe-wide: in 1878, Berlin abandoned chemical treatments to follow Paris's example. By 1900, crop irrigation sites covered 6800 ha and gave employment to 3000 farmers (*ibid.*).

Suburban agriculture is a particular phase in the history of cities' food supply systems and an innovative production model that is closely linked to urban life. It made possible great technical feats, as illustrated by the market gardeners and fruit growers of Argenteuil and Montreuil. Remarkable results were achieved by '*laitiers nourrisseurs*' (urban dairymen) considering that all of their cows' feed was purchased. "While farms constituted an independent economic unit, these suburban operations undertook only one phase of production" (Philipponneau 1952: 204).

It is true that this virtuous circle of urban ecology, whereby urban waste such as manure and wastewater was reused as fertilizer in periurban agriculture, fell into decline in the twentieth century. After the Second World War, suburban farms closed and farms in general became dependent, not on cities, but on distant farms and factories that provided seeds, feed, and chemical inputs. Paris's belt of market gardens disappeared due to such forces as the rise of the automobile, competition for land in peri-urban areas, and the ability to quickly bring in by train fruit and vegetables from regions with a warmer climate (Stanhill 1976). From that point on, municipal policies focused on the storage and treatment of urban waste.

Conclusion

The history of urban food policy is marked by very varied objectives as well as ways and means: ancient cities of the Mediterranean that guaranteed much of the urban population access to food thanks to tribute from the colonies; medieval and modern cities that made sure of their supplies by setting up exclusive catchment areas and by financing trade; and hygiene-conscious contemporary cities that issued regulations on the use of waste and where animals could be kept.

That history, however, reflects two major shifts. First, cities had to become entities that could really pursue a food supply policy. And second, the place the city would assign to agriculture, and more broadly to all living things, had to be established.

Through a slow evolution, beginning at the end of the Middle Ages, cities gradually lost their ability to conduct food policies. The rise of the States and their struggle to establish a unified national market were decisive. Right from the beginning of the twentieth century, but especially after the First World War, States increasingly gained a stranglehold on food supplies.

Again, the rapid urbanization attendant on the Industrial Revolution frequently created headless communities unable to implement policies throughout their territory. Conurbation phenomena gave rise to 'clots' of human habitation split between a number of politico-administrative entities. As much as 50 years ago Lewis Mumford noted: "Such urban clots could and did expand a hundred times without acquiring more than a shadow of the institutions that characterize a city in the mature ecological sense" (Mumford 1989: 458). He went on to say:

"The new urban emergent, the coal-agglomeration, which Patrick Geddes called the conurbation, was neither isolated in the country nor attached to an old historic core. It spread in a mass of relatively even density over scores or even hundreds of square miles. There were no effective centres in this urban massing: no institutions capable of uniting its members into an active city life: no political organization capable of unifying its common activities" (Mumford 1989: 470).

How can policies exist, absent any entity able to design and implement them? That is the very issue which, in modern Europe, is leading to the creation of metropolises that merge a number of cities into one.

Cities' separation from the organic world is the second long-drawn-out shift. This more recent phenomenon is also a consequence of the Industrial Revolution. Unlike the ambition nourished by Ildefons Cerdà, who wanted to "ruralize the city, urbanize the country" in his 1859 plan for Barcelona (Consales 2004: 798), no such synthesis occurred. While the countryside did become increasingly urbanized, thanks to the automobile, the reverse never came to pass. Cities' close relationship with the rural and organic world—through the presence of animals and market gardening, and organic processing activities such as slaughterhouses or tanneries, and through the use of urban waste in agricultural production—has disappeared in favour of simple importation of food products, lacking any connection to the living world.

These slow transformations are today being called into question, as new political spaces are being created so that cities can again be involved in food issues. At the same time, States' policies are being challenged by liberalization. Metropolises that can endow today's conurbations with a capacity for action commensurate with their size are being created. Lastly, the dispersal of production chains is increasingly being challenged on account of the risks it entails for the environment and health.

Food supply is once more on cities' agenda (Chapter "Cities' Strategies for Sustainable Food and the Levers They Mobilize"), and it follows that urban food policies can no longer discount agricultural production and the rural world. In their struggle against the cities, States often allied themselves with the countrysides, as appears from the protection policy adopted in France after the Commune by Méline and the self-sufficiency policies of the twentieth century. Today, one stage in devising ambitious urban food policies is certainly the building of new alliances with agricultural and rural areas.

References

Atkins PJ (2012) Animal cities: beastly urban histories. In: Farnham Surrey. Ashgate Publishing Ltd., Burlington

Aymard M (1966) Venise, Raguse et le commerce du blé pendant la seconde moitié du 16ᵉ siècle. SEVPEN, Paris

Aymard M (1983) Autoconsommation et marchés: Chayanov, Labrousse ou Le Roy Ladurie? Annales. Histoire, Sciences Sociales 38(6):1392–1410

Bairoch P (1985) De Jéricho à Mexico, villes et économie dans l'histoire. Gallimard, Paris

Bairoch P (1993) Mythes et paradoxes de l'histoire économique. La Découverte, Paris

Barles S (2005a) L'invention des déchets urbains: France, 1790–1970. Seyssel, Champ Vallon

Barles S (2005b) Experts contre experts: les champs d'épandage de la ville de Paris dans les années 1870. Histoire urbaine 3:65–80

Barles S (2012) Undesirable nature: animals, resources and urban nuisance. In: Atkins PJ (ed) Nineteenth century Paris. Animal cities: beastly urban histories. Ashgate Publishing Limited, Farnham, pp 173–187

Braudel F (1984a) Civilization and capitalism, 15th to 18th century. In: The perspective of the world, vol 3. Collins, London

Braudel F (1984b) Civilization and capitalism, 15th to 18th century. In: The structures of everyday life: the limits of the possible, vol 1. Collins, London

Braudel F (1995) (1949) The Mediterranean and the Mediterranean world in the age of Philip II, vol 1. University of California Press, Berkeley/Los Angeles

Chandler T, Fox G (2013) 3000 years of urban growth. Elsevier, Burlington

Clément A (1999) Nourrir le peuple – Entre État et marché, xvIᵉ-xIxᵉ siècles. L'Harmattan, Paris

Clubbe J (1992) Cincinnati observed: architecture and history. Ohio State Univ Pr, Columbus

Consales J-N (2004) Città e ambiente – Les jardins familiaux dans l'arc méditerranéen: cent ans d'agriculture dans la ville. Analyse comparative de trois agglomérations: Marseille, Gênes et Barcelone. Mélanges de l'école française de Rome 116(2):1000–1011

Frioux S, Chauveau S, Fournier P (2011) Hygiène et santé en Europe de la fin du xvIIIᵉ siècle aux lendemains de la Première Guerre mondiale. University of California Press, Berkeley/Los Angeles

Grantham GW, Sarget MN (1997) Espaces privilégiés: Productivité agraire et zones d'approvisionnement des villes dans l'Europe préindustrielle. Annales. Histoire, Sciences Sociales 52:695–725

Guerrand RH (1985) Les Lieux: Histoire des Commodités. Éditions La Découverte, Paris

Heckscher EF (1936) Revisions in economic history: V. Mercantilism. Econ Hist Rev 7(1):44–54

Heckscher EF (2013) (1935) Mercantilism. Routledge, London

Hibbert A (1963) The economic policies of towns. In: Postan M, Rich EE, Miller E (eds) The Cambridge economic history of Europe, vol 3. Cambridge University Press, Cambridge, pp 157–229

Hohenberg PM, Lees LH (1995) The making of urban Europe, 1000–1994. Harvard University Press, Cambridge

Jorland G (2013) L'hygiène publique, fille des Lumières. Tribunes de la Sante 1:23–27

Keene D (1998) Feeding medieval European cities, 600–1500. In: E-seminars in history. Institute of Historical Research, London, p 6

Lapeyre H (1969) L'Organisation municipale de la ville de Valence (Espagne) aux xvie et xviie siècles. Villes de l'Europe méditerranéenne et de l'Europe occidentale du Moyen Âge au xixe siècle. Annales de la Faculté des Lettres et Sciences Humaines de Nice Nice:9–10

Marin B, Ventura P (2004) Les Offices "populaires" du gouvernement municipal de Naples à l'époque moderne. Premières réflexions. Mélanges Casa de Velázquez 34(2):115–139

Marin B, Virlouvet C (2003) Nourrir les cités en Méditerranée – Antiquités – Temps modernes. Maison Méditerranéenne des Sciences de l'Homme, Maison & Larose, Paris

Martinat M (1999) Le Blé du pape. The Annona system in modern times. Annales 1:219–244

McNeill WH (1976) Plagues and peoples. Anchor Press, New York

Muller S (2004) Les Abattoirs sous haute surveillance. Revue d'histoire moderne et contemporaine 51(3):104–120

Mumford L (1989) (1961) The city in history. Harcourt, London

Niwa N (2015) De l'agriculture urbaine à la toshinogyo. Une analyse de leur émergence dans le cas de Genève et de Tokyo. Doctorate, Université de Lausanne

Otter C (2005) Civilizing slaughter: the development of the British public abattoir, 1850–1910. Food Hist 3(2):29–51

Otter C (2006) The Vital City: public analysis, dairies and slaughterhouses in nineteenth-century Britain. Cult Geogr 13(4):517–537

Paquy L (2004) Santé publique, répression des fraudes et action municipale à la fin du xixe siècle: le laboratoire grenoblois d'analyses alimentaires. Revue d'Histoire Moderne & Contemporaine 51(3):44–65

Philipponneau M (1952) Les Caractères originaux de la vie rurale de banlieue. Annales de Géographie 61(325):200–211

Pirenne H (1917) Les Anciennes Démocraties des Pays-Bas. E. Flammarion, Paris

Pirenne H (1927) Les Villes du Moyen Âge: essai d'histoire économique et sociale. Bruxelles, Maurice Lamertin

Quellier J (2015) "Paris is land of plenty": kitchen gardens as urban phenomenon in a modern-era European City (sixteenth through eighteenth centuries). In: Imbert D (ed) Food and the City. Histories of culture and cultivation. Dumbarton Oaks, Washington, DC, pp 273–300

Reid D (1993) Paris sewers and Sewermen: realities and representations. Harvard University Press, Harvard

Revel J (1975) Les Privilèges d'une capitale: l'approvisionnement de Rome à l'époque moderne. Annales. Histoire, Sciences Sociales 30(2/3):563–574

Scola R, Scola P (1992) Feeding the Victorian City: the food supply of Manchester, 1770–1870. Manchester University Press, Manchester

Sen A (1981) Poverty and famines: an essay on entitlement and deprivation. Oxford University Press, Oxford

Stanhill G (1976) An urban agro-ecosystem: the example of nineteenth-century Paris. Agro-Ecosystems 3:269–284

Steel C (2008) Hungry City: how food shapes our lives. Chatto & Windus, London

Tanguy JF (2007) Le Laboratoire municipal de Rennes et l'hygiène alimentaire (1887–1914). Villes en crises? In: Marec Y (ed) Les politiques municipales face aux pathologies urbaines (fin XVIIIe-fin XXe siècle). Creaphis, Paris, pp 436–453

Taylor-Leduc S (2015) Market gardens in Paris: a circulus intelligent from 1790–1900. In: Imbert D (ed) Food and the city. Histories of culture and cultivation. Dumbarton Oaks, Washington, DC, pp 300–332

Thick M (1998) The neat house gardens: early market gardening around London. Prospect Books (UK), Totnes

Thompson F (1976) Nineteenth-century horse sense. Econ Hist Rev 29(1):60–81

Tilly C (1974) Food supply and public order in modern Europe. The formation of National States in Western Europe. Princeton University Press, Princeton, pp 380–455

Tilly C (1990) Coercion, capital, and European states, AD 990–1990. B. Blackwell, Cambridge, MA

Trench R, Hillman E (1984) London under London: A subterranean guide. Murray, London

Weber M (1991) Histoire économique: Esquisse d'une histoire universelle de l'économie et de la société. Gallimard, Paris

Cities' Strategies for Sustainable Food and the Levers They Mobilize

Jess Halliday with Wendy Mendes (Box Contributor)

There is a growing realization of cities' vulnerability to the problems posed by the food system that are described in the chapter "Urbanization Issues Affecting Food System Sustainability". Reliance on globalized supply chains puts food provisioning arrangements at risk from environmental, political or economic disruptions. Cities lack productive space to produce all the food needed to feed their populations (Steel 2008, 2012), yet cities in less developed countries of the Global South will host most of the growth in world population, projected to reach 9.1 billion by 2050 (FAO 2009). Household food insecurity (or food poverty) is a major issue in many cities in developing countries, but incidence has also risen in the Global North, where the long-term effects of the 2008 economic downturn are still felt. Moreover, urban consumers are increasingly disconnected from the origins of their food (see chapter "Urbanization Issues Affecting Food System Sustainability"), and modern cities are obesogenic environments where calorie-dense/nutritionally-lacking food is cheap and available, but where opportunities for physical activity are limited (Morgan and Sonnino 2010). Consequently, diet-related ill-health has reached epidemic levels.

While the capacity of cities to implement food policies has ebbed since the end of the Middle Ages (see chapter "History of Urban Food Policy in Europe, from the Ancient City to the Industrial City"), a small but increasing number of cities around the world are devising policies to address food-related problems or to mitigate their effects. This chapter describes the range of cities' aims in so doing, while drawing on existing literature to develop a broad typology and exploring relations between cities and their hinterlands, since many policies have repercussions that are felt beyond the boundaries of cities or involve external actors. The analysis in this chapter focuses on the levers and instruments that are employed by cities to meet their objectives, with attention to the importance of cross-domain working and determining

J. Halliday (✉)
CIRAD, UMR MOISA, Montpellier, France
e-mail: j.halliday@ruaf.org

© The Author(s) 2019
C. Brand et al. (eds.), *Designing Urban Food Policies*, Urban Agriculture,
https://doi.org/10.1007/978-3-030-13958-2_3

what is possible within each city's context. A typology of governance models for cities' food-related interventions is then outlined, and the need to draw on capabilities of civil society and the private sector to overcome barriers faced by the public sector is highlighted. Several ways in which the success of urban actions can be measured are subsequently discussed. Finally, two possible future directions for cities' role and influence in the food system are identified.

Objectives of Urban Food Policies and Challenges of Urban-Rural Relationships

Any attempt to find homogeneity in the intentions behind cities' food policy efforts is fraught with difficulty.

Firstly, political and socioeconomic histories have led to differences in perspectives and priorities with regard to food. These differences are apparent from city to city but are most evident when comparing cities in developed and less developed countries. In many places in the Global South, food security is supported by close connections between rural and periurban producers and urban consumers, and (often informal) urban agriculture, but migration to cities means connections are failing and encroaching urbanization causes land-use conflicts. Prosperous cities of the Global North, on the other hand, tend to rely heavily on the globalized, industrial food system that transports ingredients and composite products across continents. Yet in developed countries there are wide inequalities in access to healthy, nutritious and quality food. Food insecurity is often manifested as obesity rather than starvation, while 'precarious food situations', as defined by Dominique Paturel (Paturel et al. 2015), are marked by a lack of physical, economic or cognitive access to healthy and nutritious food.

Secondly, there is a lack of consistency in how cities express their aims, their entry points and framing of the issues. It has been proposed that cities' main aim is to ensure future food security (Morgan and Sonnino 2010; Sonnino 2014), which is a situation when, "[...] all people, at all times, have physical and economic access to sufficient safe and nutritious food that meets their dietary needs and food preferences for an active and healthy life" (World Food Summit 1996). The phrase 'food security' seldom appears in the titles of urban food strategy documents in the Global North, as Roberta Sonnino (2014) found in her discourse analysis of 15 such strategies, but it tends to be implicit and embedded within health and sustainability language. While there is sometimes direct reference to food security in the body of these documents as one of several priority themes, it is usually framed as a health or food access issue.

Thirdly, the aim of a food-related policy is not always articulated separately from the levers to be employed to implement it. Sometimes a city's entry point is not the identification of a problem to be addressed, but rather an aim, a strategy, or a desire to implement an available instrument (as discussed in greater detail in chapter "Reconciling Sustainability Issues and Urban Policy Levers").

Fourthly, not all food-related policy is framed as such, but the headline aim can be wider, with food being one of several strands. This is the case for Medellin's *Todos por la vida* ('everyone for life') policy, which builds the city's infrastructure to alleviate the effects of the civil war, with an ethos of greening, biodiversity and social inclusion. The wider programme includes food security enhancement strategies (Laidlaw 2015).

Despite these difficulties, nine food-related aims were identified in a review of literature on cities' food policies published in 2015 or earlier.[1] These aims are presented in Table 1, with examples of policy levers and instruments through which 11 cities work towards their aims. Levers are defined as policy domains or responsibilities that reside at the local level and which usually have an associated local government department or service. Instruments are the procedural means that the departments or services have at their disposal (see chapter "Reconciling Sustainability Issues and Urban Policy Levers").

Regeneration, the first aim in Table 1, uses food-related activities to revitalize the social fabric of an urban area that has become run-down following a crisis. In Detroit, urban agriculture could be a means of regenerating the city after its economic collapse, as Detroit is exempt from the State Right to Farm Act which otherwise prohibits agriculture in urban areas (Detroit Future City 2012). In Medellin, the crisis was sociopolitical, and funding was provided for urban agriculture projects by the State electricity provider as part of a long-running urban development programme (Laidlaw 2015). In both of these examples, urban agriculture is central to regeneration efforts and can address problems of physical or economic access to healthy food, while boosting cognitive reconnection with food sources and food production, thereby helping to improve diets.

Regeneration is closely connected to economic development, the second aim, as diverse and prosperous food businesses reinvigorate the economy through their own activities and by attracting customers for all local businesses. Within cities, viable food enterprises contribute to local economic prosperity through taxable revenues and job creation, with an emphasis on *good* jobs that pay a living wage and offer safe and fair conditions and training in healthy food preparation and sustainability (Freudenberg and Silver 2013). In London, funding and campaigning for food apprenticeships by the Greater London Authority helps tackle joblessness whilst creating a pool of skilled professionals (Johnson 2013). Moreover, as in Barcelona, investment in food market infrastructure yields benefits within the city and regionally. It improves food access in cities, while securing access to the profitable urban market for regional producers, thereby helping to overcome long-term resilience problems by ensuring that the local food supply is viable (Forster et al. 2015).

---

[1] Academic papers were identified through a search of City University London's library online using the terms 'food policy' and 'city' or 'urban', and 'food policy council', and 'urban food strategy'. Non-academic literature was identified through online searches (Google) using the same terms. Policy documents were sourced where the author had prior awareness of food policies. The policies included in Table 1 are examples and not a definitive list of cities' food-related policies.

Table 1 Examples of cities' aims, levers and instruments

Aims	Levers	Instruments
1. Regeneration		
Detroit, USA	Planning	Urban agriculture zoning regulation
Medellin, Colombia	Urban development	Funding urban agriculture
2. Economic development		
Barcelona, Spain	Market management	Investment in food market infrastructure
London, UK	Economic development	Campaigning, funding food apprenticeships
3. Food supply resilience		
Belo Horizonte, Brazil	Agricultural development	Technical and financial incentives for agriculture; Production-consumption schemes; Promoting urban agriculture
4. Food security and access		
Belo Horizonte, Brazil	Social welfare	Subsidies for canteens and staple foods; School food funding
Medellin, Colombia	Business/trade Social welfare	Price regulation; Public-private partnerships; Participatory budgeting for emergency food
Toronto, Canada	Urban development Public health	Funding for urban agriculture; Subsidies for mobile fruit and vegetable vending
5. Environmental protection		
Malmö, Sweden	Public canteens	Procurement contracts; Training
London, UK	London 2012 Olympics	Contracts specifying sustainability criteria
6. Public health		
New York, USA	Public health Public procurement	Regulation on trans fats, calories on menus; Executive order on nutrition standards
Waltham Forest, UK	Planning	Planning restriction on food takeaways
7. Food safety		
Hanoi, Vietnam	Retail modernization	Public-private partnerships; Private standards
8. Social inclusion		
Rosario, Argentina	Planning	Public land use designation for urban agriculture; Tax relief for private landowners; Participatory budgeting
9. Food culture		
London UK[a]	Mayoral support	Funding on urban agriculture; Support campaign

[a]Social and cultural aspects of London food is one of several themes in the London Food Strategy, while the others are covered under other identified headings

The next two aims, food supply resilience and food security and access, are also connected. Resilience is a precursor to food security as it ensures that the food system can withstand macro-level threats, be they economic, sociopolitical or environmental. The example of Belo Horizonte bears similarities to the market infrastructure development policy in Barcelona, since agricultural development is intended to ensure local production will be able to meet demand in the long term. Food security and access means ensuring that nutritious food is available to the entire community. In Medellin, policy efforts include the provision of emergency food aid under social welfare, and price regulation (Laidlaw 2015). In Belo Horizonte, social welfare efforts are focused on subsidizing canteens and funding school meals (Rocha and Lessa 2009). Physical access can be improved by bringing food supply into underserved areas, such as Medellin's public-private partnerships to ensure that businesses operate in food deserts (Laidlaw 2015). In Toronto, physical access is addressed through integrating urban agriculture into residential areas, and via subsidizing vans to deliver fresh vegetables and fruit in underserved neighbourhoods (Mah and Thang 2013).

The environmental protection aim also promotes long-term resilience as any harmful environmental impacts of food production, distribution and waste disposal can impair future food security. Environmental problems hinge largely on the interface between production and demand within cities. On the one hand, protecting the environment comes down to informing producers about agricultural technologies that are more sustainable and encouraging their adoption. On the other hand, there is a need to create demand for food produced using environmentally-sound practices, as achieved in both Malmö (Anderson and Nillson 2012) and London (London, 2012 2009), where agricultural production practices are specified in procurement and catering contracts.

The public health improvement aim is intended to directly address problems caused by long-term consumption of inappropriate foods, as opposed to the food safety aspects of public health discussed below. Public health is linked to the food security aim, as optimum public health is an outcome of access to sufficient, safe and nutritious food, but enabling public is not just about preventing malnutrition. The ubiquity of cheap, convenient food of poor nutritional quality is a driver of the obesity crisis and diet-related disease (Morgan and Sonnino 2010). Public health measures tend to fall into two categories: education and consumer awareness campaigns to enable informed food choices; and altering the food environment to promote healthier options or limit unhealthy options. The requirement of calorie labelling on restaurant menus in New York is a prime example of the first category, while the second is illustrated by city Health Code amendments prohibiting artificial trans fats in takeaway foods (Libman 2015). Planning restrictions on the opening of new takeaway restaurants near schools and playgrounds in the London Borough of Waltham Forest is a further example of a change in the physical environment (GLA and CIEH 2012). Food safety is a basic facet of public health, although it can also be a specific aim geared towards avoiding the immediate, and often catastrophic, effects of poor hygiene or lack of infrastructure on human health. This is the case with traditional wet markets in Hanoi (Vietnam) that were perceived to pose a food safety hazard since they do not meet the private standards of modern supermarkets. Consequently, the city's retail modernization policy framework, which favours public-private partnerships, has led to their closure and replacement with supermarkets (Wertheim-Heck et al. 2015).

Social inclusion and food culture—the last two aims—seek to remedy the problem of consumers' cognitive and cultural detachment from food. Social inclusion concerns the social and participatory nature of eating, preparing or growing food. It strives to deal with inequality, such as in Rosario, where the planning system is used to provide access to public or private land for those who would otherwise be excluded from food growing (Roitman and Bifarello 2010). Finally, the food culture aim seeks to foster conscious food choices. In London, Mayoral support has enabled programmes to educate the public and raise awareness about sustainability and healthy food. Events are organized to encourage food diversity, from both nutritional and cultural standpoints, particularly in light of the ethnic diversity within the city (GLA 2011).

Each of the aims in Table 1 ostensibly seeks to shape the food environment within the city for the benefit of urban residents. However, several of them have effects or require action outside of the city limits too—notably economic development, food supply resilience, food security and access, and environmental protection. This is because the food system is an intrinsic and non-divisible whole made up of all stages of food supply, as well as contextual influences, inputs, outcomes and outputs (Lang et al. 2009; Lang and Heasman 2015). Moreover, as the food system is not geographically bounded, all places of human settlement, their demands and impacts have knock-on effects throughout the system, worldwide (Marsden 2013). Despite this, urban and rural affairs historically have been regarded as discrete disciplines, with the former being a locus of consumption and the latter of production (Dupuis and Goodman 2005). As local administrative boundaries often stop at the city limits, further discussion on the links between cities and periurban and rural areas is warranted.

The City Region Food System (CRFS) concept is an attempt to bridge the long-standing divide between urban and rural concerns (Dupuis and Goodman 2005; Cohen and Garrett 2010). In 2013, CRFSs were defined as encompassing:

> [...] a complex network of actors, processes and relationships to do with food production, processing, marketing and consumption that exist in a given geographical region that includes a more or less concentrated urban centre and its surrounding periurban and rural hinterland; a regional landscape across which flows of people, goods and ecosystem services are managed (www.cityregionfoodsystems.org).[2]

Local and regional provisioning is an operational facet linking urban consumers with rural producers via food value chains, but the intention is wider—to maximize ecological and socioeconomic links and to foster food system co-governance by both urban and regional actors (Jennings et al. 2015). In this way, CRFSs avoid falling into the 'local trap' that assumes local is an inherently good scale for action,

---

[2] Definition by a collaborative partnership of interested organizations.

either by considering localisation of supply chains as an aim in itself or by failing to explore the role of other scales in configuring urban food environments (Libman 2015; Born and Purcell 2006; Winter 2003).

However, while CRFS is a helpful concept that is being applied in empirical settings, its applicability is nonetheless contingent on two factors: the situation of the city within an agricultural basin; and the existence of regional institutional structures and their willingness to engage with the city.

Firstly, according to the definition, a 'city region' comprises an urban centre and a periurban and rural hinterland, with provisioning links dependent on the latter being a productive agricultural basin. The Dutch city of Rotterdam, with arable, dairy, and greenhouse horticulture production in the surrounding province of South Holland, is a good example of a CRFS (Van der Schans 2015). On the other hand, La Paz (Bolivia), where the high elevation and low temperatures limit agricultural production and most provisions are brought in from elsewhere (Velasco and De Vrieze 2015), cannot be considered as a CRFS centre. Indeed, Braudel (1979) points out that, historically, cities were not systematically established at the centre of a productive region that could serve them, but rather access to trade routes—usually via a port—was crucial to the most powerful cities, allowing them to draw provisions from elsewhere.

Secondly, co-governance between urban and regional actors can depend on the existence of governmental and institutional structures at the regional level, and political will within them. In the UK, England's regional layer of government was dismantled by the coalition government that came to power in 2010. The Netherlands, on the other hand, is divided into 12 regions with clear policy responsibilities, providing a framework for cities to engage with the wider region on food issues. In France, meanwhile, the regions have expressed an interest in more territorialized food systems and have suggested instruments for regional authorities to contribute to sustainable food regions, such as regional marketing labels and technical or financial support (ARF 2014; Braine-Supkova and Gaspard 2015). This does not make regions automatic vehicles for coordinated cross-boundary food policy, however. At the metropolitan level, meanwhile, there is considerable potential for cooperation across administrative boundaries. In the *Montpellier Méditerranée Métropole*, 31 municipalities are cooperating to develop a food strategy (see chapter "Reconciling Sustainability Issues and Urban Policy Levers"). Similarly, in England, the emergence of Greater Manchester as a pilot city region (following the demise of the English regions) may enable integrated food policy implementation across a larger area (Halliday 2015).

Urban Food Policy Levers and Constraints on Their Use

Alongside the broadly categorized aims, Table 1 also sets out levers and instruments employed by cities in pursuit of their aims.

The examples of Belo Horizonte, Medellin, Toronto and New York show that the levers and instruments to address an aim are not always limited to one policy area, but several can be employed together. This is because food is not the sole responsibility of a single local government department but different aspects are directly or indirectly handled by different teams (such as public health, economic development, planning, education, etc.). On the one hand, this means there is potential for a more powerful multi-pronged effort, such as in Toronto where food access interventions (e.g. subsidized mobile fruit and vegetable vans) are complemented by public health efforts to boost food literacy, and as a consequence demand for healthy food (Mah and Thang 2013). On the other hand, unless there is recognition of the cross-domain nature of food and a concerted effort towards coordination, there is a danger that each department will look no further than the implications on its own remit (Pothukuchi and Kaufman 1999).

Recognition of the cross-domain nature of food is increasingly seen as the best practice, although it is not yet ubiquitous. Coline Perrin and Christophe Soulard (2014) and Caroline Brand (2015) report that in France to date there is little recognition that food is a local policy area per se, or that food issues require concerted integrated action at the local level. A common intention of food policy councils[3] that have emerged over the last 20 years in the Global North (particularly in English-speaking countries) is the intention to find and exploit synergies between different public sector areas involved with food (Wiskerke and Vilojoen 2012). One area often serves as a hook or entry point to gain initial local government buy-in in keeping with civic priorities. The initial framing of both the Toronto Food Policy Council in Canada and the London Food Board in the UK was public health but both maintained wider underlying policy objectives on food access, affordability, education and production. This undercurrent of cross-domain activity prevents issues being overlooked and enables them to be brought to the fore if priorities change (Blay-Palmer 2009; Mah and Thang 2013; Reynolds 2009; Morgan and Sonnino 2010).

These examples and best intentions notwithstanding, achieving synergies between policy areas is not always easy. Firstly, each area or department has its own procedures and ways of working (tacitly or explicitly agreed) and language that may not be understood in co-working situations. Departments do not always recognize the relevance and value of food to their core work, nor that food is a cross domain issue over which there is much potential for inter-departmental working (Wegener et al. 2013). Rigid line management structures can also limit the potential for senior staff to take part in strategic policy work with other departments (Yeatman 2003).

Secondly, the precise distribution of powers and responsibilities at the local level varies between national settings. The principle of subsidiarity (which has become a fundamental element in the functioning of the European Union since the early 1990s) holds that policies should be formulated and enacted at the most local level possible and higher levels should step in only if the required action cannot be

---

[3] Morgan (2014) counted 193 food policy councils in North America in 2012, while in the UK the term 'food partnership' is more prominent and some 40 food partnerships have been founded in local government areas in recent years.

achieved. This seems to give local governments *carte blanche* to devise food policies as they see fit. However, the EU protocol on applying the principle of subsidiarity relates to EU institutions and Member States (European Union 2010), while there are only recommendations on the division of responsibilities between national and sub-national levels whereby each state can devise its own system (Council of Europe 1995). No ideal level is prescribed for each policy area because local authorities vary in size and resources.

Moreover, the role of local government and its leaders with regard to the national level differs between countries. In some, local government has full legislative power in locally-managed domains; in others it has the power to introduce a policy that is not the specific responsibility of another level or entity; while in others local government largely delivers services or commissions on behalf of the national level. A comparison of the powers of the mayors of New York and London with regard to healthy eating illustrates these differences. The Mayor of New York has regulatory powers that have enabled him to introduce restrictions on trans fats in fast food establishments and calorie labelling in chain restaurants (Libman 2015). His London counterpart has no legislative power and his public health remit is limited largely to advocating conducive policies at the borough level,[4] and seeking voluntary measures (Halliday 2015; Morgan and Sonnino 2010). In France, cities lack the political weight of their European and American counterparts, with responsibility for food residing at the national or EU level (Perrin and Soulard 2014). Meanwhile, in North African countries, agricultural policies for food security are national rather than local (Soulard et al. 2015).

A third problem in the use of public sector levers is that options within various policy areas are subject to multilevel governance constraints. European public procurement policy is a prime example: while procurement for all locally-managed public services takes place locally, the EU public procurement directive stipulates that calls for tender may not specify preferred place of origin or use the word 'local' (Morgan and Sonnino 2008). Planning policy in England is a further example: each local government authority produces its own planning strategy which must be in keeping with the National Planning Policy Framework (NPPF). Consequently, although the London Borough of Waltham Forest introduced restrictions on new unhealthy fast food outlets near schools and playgrounds, NPPF removed the need for prospective fast food outlets to obtain permission to change the use of business premises, thus making Waltham Forests' new policy hard to implement. Such barriers highlight the need for local government to engage with multiple levels and advocate conducive framing policies.

In light of the arbitrary application of the principle of subsidiarity, differing roles for local government with regard to the national level, and multilevel framing constraints, it is not possible to definitively identify which policy areas are best handled at which level. Planning and public procurement have been identified as two areas in which the local level has the greatest potential to address food issues (Sonnino 2014; Sonnino and Spayde 2014), but there are other policy areas related to food

---

[4]London has 32 boroughs, each with its own local government.

Table 2 Food-related policy areas

Policy area	Relevance to food
Planning, infrastructure and urban development	Land use, location of food outlets/retail
Public transport and roads	Infrastructure for food distribution and retail market access
Local economic development	Supporting food businesses, job creation, development of skills
Local finance administration	Tax on food businesses, budgetary provision for services and subsidies
Social welfare services	Food aid (food vouchers and subsidies), emergency food provision
Public health	Diet-related preventive health, education and targeted campaigns
Education	Food skills and culture
Public procurement	Food served in public canteens and institutions
Environmental sustainability	CO_2 emissions from production and distribution, natural resource use, biodiversity preservation
Environmental health	Implementation of food safety regulations, environmental conditions
Waste management	Redistribution of surplus food, composting and recycling

where the local level may have a role, be it devising policy, interpreting/implementing policy from higher levels, or service provision. These are set out in Table 2.

Actors in each city are advised to reflect upon the applicable national model for local government and policy areas that can be leveraged, while acknowledging the constraints and barriers faced. By being realistic, they may find ways to remove the barriers (e.g. through negotiation, relationship building and subtle influence) or, where barriers are immutable, to identify creative workarounds to achieve the same results. These workarounds can include drawing on the resources, levers and instruments of civil society and private sector actors who support the objectives, but who are not subject to the same institutional constraints. The prospect of non-public sector levers and instruments calls for an analysis of the governance of city food policy initiatives.

Involvement of Different Sectors and Respective Governance Models

While so far this chapter has focused on public policy to address food-related problems, this section discusses the involvement of civil society and the private sector. Three governance models, each with variations, are outlined. Over the last 30 years, the shift to governance (particularly in developed countries) has brought the private sector and civil society into the policy-making process. This should not be interpreted as meaning that the State is handing over power (Pierre and Peters 2003), but rather that the State no longer 'rows' (i.e. delivers services or makes decisions unilaterally) but instead 'steers' using tools such as monitoring, cultural persuasion, financing, and a reserved right to intervene (Stoker 2000). This shift has taken place in every policy domain and means that the food system overall is not governed solely by top-down regulations. It is subject to contested governance in which the public and private sectors and civil society cooperate, but also compete for influence. However, the three sectors are not always perfectly represented when it comes to food policy development and implementation at the local level.

There is general agreement that political buy-in to food policy at the local level is critical to achieve change (Orlando 2011; Schiff 2008). Many commentators have nevertheless highlighted the benefits of civil society working alongside local governments. In non-Anglo Saxon contexts, Perrin and Soulard (2014) underline the need to conduct studies on forms of civil society involvement in food policy and to assess their contribution. If a food policy council only involves the public sector with no civil society ballast, there is a danger it will be unable to propose policy changes to the system within which it operates. As noted above, civil society and private sector actors bring their own arsenals of resources and instruments to complement those of the public sector. Using these, they can exercise agency, i.e. "act consciously and, in so doing to attempt to realise his or her intentions" (Hay 2002, p. 94), so as to overcome barriers posed by public sector institutional frameworks, e.g. inhospitable policy framing in multilevel governance or immutable procedures and received language (Halliday 2015). Civil society involvement is a key principle of food policy councils, which are groups of actors from multiple sectors. The latter meet regularly and their activities include formulating and advocating policy approaches, raising awareness, knowledge brokerage, and monitoring progress. Civil society brings insight and energy (Derkzen and Morgan 2012), offers resilience in the face of electoral change (Wiskerke 2009; Morgan and Sonnino 2008), and can provide specialist food knowledge and experience (Wekerle 2004).

As for the private sector, almost all food passes through the hands of at least one private enterprise on its route from producer to consumer, yet Harper et al. (2009) point out that food businesses are often un- or under-represented in food policy groups, and suggest that if they are not engaged from the start it is difficult to bring them in later if decisions have already been made that are not seen to be in their interest. This poor representation of food businesses could be explained by the perception that local food policy is entirely opposed to the global food system operations, with aims and objectives that are incongruent with those of business. However, a change in business practices has repercussions throughout the supply chain. Hence, local food policy groups should ideally include diverse businesses of different sizes, values, and activities to be able to initiate effective system change (Halliday 2015).

Food policy councils are the first governance model for local food policy. As a minimum they include actors from the public and civil society sectors, but there is much internal diversity in how they are initiated and run. Some, such as Manchester Food Futures in the UK, were initiated via top-down processes, while others, like the London Food Programme, were formed through a combination of top-down and bottom-up energy (Reynolds 2009). Where initiation is top-down, questions remain about when civil society and the private sector are to be brought in, who decides which organizations are eligible and on what criteria. If they join too late, they may struggle to achieve ownership of a preset agenda. The institutional positioning of food policy councils also varies. Some are hosted within local government structures, e.g. the Toronto Food Policy Council is embedded within the public health team (Blay-Palmer 2009). Others occupy an independent space outside local government, like the Bristol Food Policy Council which, nonetheless, still involves public sector actors (Carey 2013). Finally, the nature and degree of civic leaders' commitment ranges from explicit mayoral support (as in London), to adoption of a food strategy or motion (as in Manchester), to providing budget or in-kind support such as meeting rooms or officers' participation during working hours.

Despite the popularity of the food policy council model, it is by no means the only type of urban food policy intervention. A second type is found in some cities, where the intervention is formally rooted in a public mandate to take responsibility for the food system, albeit with input or practical contributions from civil society, the private sector, or universities and research institutes. In San Francisco in 2009, the then-Mayor Gavin Newsome issued an executive directive making the food system the explicit responsibility of the city government. In La Paz (Bolivia), the 2014 'Autonomous Municipal Act N°105 on Food Security', the goal of which is to secure the right to food for all citizens, was prompted by campaigning by civil society groups. The benefit of high level mayoral commitment is that it enables (and often obliges) departments and agencies to review their food-related policies. Elsewhere, a local government entity has been set up to coordinate integration between food-related policy domains, as is the case with the Municipal Secretariat for Food Policy and Supply in Belo Horizonte, which engages the private sector and civil society over implementation (Rocha and Lessa 2009). As for universities, the development of a city-region agriculture and food production policy in Montpellier (France) has involved a team of researchers working alongside the public sector (Soulard et al. 2015; chapter "Reconciling Sustainability Issues and Urban Policy Levers"), while researchers from the University of Pisa (Italy), with the agreement of the provincial administration, initiated multisector discussions over developing a food plan for the city (Di Iacovo et al. 2013).

While a coordinating entity is helpful, it is not a prerequisite of interventions that have positive impacts across policy areas. For instance, the city council in Ourense (Spain) initiated an agricultural land-use policy of establishing urban orchards. This initiative was intended primarily to boost citizens' education and employment skills using financial instruments to facilitate access to private land, but it has also brought documented environmental, social and economic benefits (Oursense City Council 2013). The London borough of Waltham Forest's planning restrictions on unhealthy fast food outlets, as mentioned above, brings both public health and environmental benefits.

A third type of urban food policy initiative that is worthy of mention consists of civil society interventions with varying degrees of organization. In some cases, food activities are part of a wider social movement, such as Transition Towns through

Table 3 Overall types and variations of urban food policy interventions

Types	Variations
Food policy council	Top-down or bottom-up initiatives (or both) Hosted within local government or outside Nature and extent of local government support
Formal public-sector mandate	Executive directive or law Local government entity to coordinate integration of different areas Individual policies with cross-policy domain implications
Civil society intervention	Part of a wider social movement Set up by an established organization with funding Guerrilla movement

which grassroots communities seek to build resilience in response to peak oil, climate change, and economic instability (Morgan 2009). In other cases, food projects are instigated by established organizations, usually upon receipt of funding. While such projects initially lack local government support, they might clinch it once they are established and yielding positive results. For instance, in Porto Alegre (Brazil), the bottom-up guerrilla gardening initiative of planting fruit trees has been accepted and maintained by municipal workers (Abelman 2015). This case is reminiscent of Incredible Edible Todmorden in the UK, which started off as community movement and has earned the support of public bodies while serving as a template for integrating agriculture into city environments around the world (Warhurst 2012).

The three types of urban food policy described above and their variations are summarized in Table 3.

The governance model for an urban food policy initiative often does not come down to a choice by the initiator, and it would be inappropriate to recommend a one size fits all model to be universally applied to all places and circumstances. Rather, different models are suitable and possible in different settings, depending on variables such as local government structure, political will, and social capital levels. The high social capital levels in Bristol (UK) have enabled the Bristol Food Policy Council to secure some influence over policy making despite residing outside of local government. Meanwhile, in Manchester—a city with a relatively conservative culture—the Food Futures programme could not have operated effectively had it not been embedded within the Manchester City Council (Halliday 2015).

Ways to Assess the Success of Urban Food Policies

The success of an intervention will ultimately need to be assessed irrespective of the governance model, aims, levers and instruments employed. This means coming up with a definition of the success of a food policy. However, it is hard to judge the impact of a policy initiative unless indicators and evaluation criteria are defined at the start (Sonnino 2013; Burgan and Winne 2012), especially because food-related issues are complex and there can be non-food contributing causes or mitigating factors. For example, if the aim is to improve public health or reduce negative environmental impacts using public procurement as a lever, a measureable indicator could be the number of school canteens or other public bodies subscribed to a certification scheme. If the aim is to improve the food economy using more local food sourcing as a lever, indicators could be the land area devoted to food production, profits of local food enterprises, or their ability to remain in business. A problem with indicators is that they are retrospective and require a period of time before success can be measured, at which point it could be too late to deal with any barriers. One way of overcoming this problem is to assess the process of establishing a governance body, while determining its role and mission and implementation as it is taking place (Burgan and Winne 2012).

Kenneth Dahlberg (1995) also offered a different interpretation of success by asking, "has everything been done that was possible in that context?". This highlights that the impact of policy efforts is dependent upon circumstances and subject to constraints within the local policy environment and in multilevel framing of policy domains. It is prescient of the strategic urban governance capacity (SUGC) approach employed by Wendy Mendes (2008) and Brent Mansfield (Mansfield and Mendes 2013) to explore factors that can affect the capacity of local government to develop and implement a food strategy (see this chapter, Box 1).

Another way of judging the success of an initiative is to assess its wider impact on the food system, with scaling up leading to reconfiguration of food provisioning arrangements and impacts outside the boundaries of the city or its immediate surroundings. Given the current enthusiasm for urban food policy and CRFS, there is a danger of giving the impression that food system problems can be solved by cities alone. Roberta Sonnino (2014) has underlined the ambitious claims by several cities of 'world leadership' with regard to urban food strategies in the Global North, which imply not only that they are among the first to have a comprehensive food strategy but also that they have the means to bring about larger-scale food system change. Actually, inhospitable multilevel governance can present barriers to larger-scale change (Barling et al. 2002) and can hamper the ability of an initiative to flourish in the long term (Morgan et al. 2006). By and large, multilevel food policy remains predominantly neoliberal, albeit with highly contested governance. To date there is little evidence of larger scale change prompted by any city's food-related policy, with two notable exceptions: Belo Horizonte's *Fome Zero* (zero hunger) strategy, which was instrumental in the institutionalization of the right to food at the national level (Rocha and Lessa 2009); and New York's trans fats directive, which prompted the Food and Drug Administration (FDA) to propose new rules that would eliminate trans fats in the US food supply (Libman et al. 2015).

Possible pathways to higher level influence include the involvement in local level food policy of actors who have political influence at other levels (e.g. public figures or big business representatives), and greater cooperation between cities. There is already evidence of cities learning from each other and sharing experiences, whilst remaining alert to the governance differences that limit direct policy transfer, as seen in the exchanges between London (UK) and Toronto (Canada) (Blay-Palmer

Box 1: Strategic Urban Governance Capacity Concept and Urban Food Policy Implementation Factors

Wendy Mendes

The strategic urban governance capacity (SUGC) concept, as theorized by urban planning scholar Patsy Healey (2002), can be traced to the shift from government to governance, resulting in a blurring of distinctions between the State, market and civil society (Harvey 1989). Beyond this commonality, SUGC can be distinguished by its focus on "the creation of meanings for the city which make a contribution to creating and sustaining an imaginative, shared collective resource, which is richer and more inclusive" (Healey 2002: 1778). In this sense, SUGC is a lens through which fundamental questions can be asked about who and what cities are for, and who has the power to shape their trajectory (*ibid.*: 1779). The SUGC concept is productive for a number of reasons with regard to urban food policies and other complex sustainability challenges.

Firstly, it shows the extent to which impulses for change must be imagined before transformations can be achieved. This means that city governance analyses must foster attentiveness not only to governmental arrangements, but equally to wider urban concepts that new social and environmental mandates such as food policy imply. Although resilient food systems are among the most ancient of urban concerns, the relatively recent re-emergence of food on urban agendas puts a spotlight on assumptions about what counts as a 'legitimate' urban system, who participates in defining food system challenges, and the ways in which solutions are co-created. This is at least partly due to the unusually high number of stakeholders associated with food movements. These include coalitions concerned with ecological protection, public health, nutrition, anti-poverty, social inclusion, community capacity-building, participatory decision making and economic development. Although not without internal tensions, such a broad landscape of perspectives can enhance dialogue—and imaginings—on contemporary city-building and civic engagement.

Secondly, the SUGC notion humanizes governmental institutions as sites of urban transformation and reveals that governance and policy making are deeply social processes (While et al. 2010; Mendes 2007, 2008; Mansfield and Mendes 2013). As Healey argues, no one agency has the power to produce the city, be it materially or symbolically (2002: 1785). An examination of governmental institutions through the lens of SUGC makes it possible to envision the potential to incubate new, sometimes radical, ideas and practices involving traditional and nontraditional stakeholders. These practices can cast light on city governance as "a border zone of trial and error" in which those involved assume new and expected roles (Appadurai 2001: 33–34).

(continued)

Box 1: (continued)

Another way to describe such an approach is what Dana O'Donovan and Noah Rimland Flower (2013) call "adaptive strategy", referring to the shift away from long held assumptions about strategy and planning as predictable, hierarchical and linear. In recognition of the complexity of our current economic, political and social systems, O'Donovan and Rimland Flower argue that the adaptive strategy allows for iterative problem solving characterized by collective decision making, rapid prototyping and experimentation (*ibid.*). Where urban food systems are concerned, the rise of large-scale urban farming, food business incubators, community food hubs, comprehensive municipal food strategies and other food-centred innovations, requires precisely this type of expanded capacity for experimentation and ability to test new solutions. Food system innovations often take place under conditions of ambiguous regulation, typically have few precedents, and involve the participation of non-traditional stakeholders. These innovations contribute to and are propelled by new understandings of governance, planning and policy making embodied in notions such as SUGC and adaptive strategy.

Thirdly, building on expanded of strategic planning notions, a SUGC approach to urban food policy disrupts institutional and sectoral silos that can impede systemic responses to complex sustainability challenges. As Healey writes, strategic planning efforts based on SUGC principles, "move from a position *above and apart*...to institutional locations *within the flow*...of urban life (Healey 2002: 1786). Results can include a challenge to centralized decision making, an increase in cross-agency collaboration, and a shake up of institutional silos (*ibid.*: 1787). For instance, in his analysis of the London food strategy, Reynolds (2009) examines the intention by the City of London to take "a holistic view of the food that the city produces, stores, delivers, sells, consumes and wastes" (*ibid.*: 417). Reynolds argues that the diversity of food issues addressed in London's strategy, combined with the wide range of agencies and sectors involved, is largely responsible for its continued relevance and success (*ibid.*: 417–418). He writes:

> [...] rather than being split into a myriad of different silos, we believe a holistic approach is necessary: where decisions are made that not only consider the healthiness of a particular food offering but also its environmental impact, where social and cultural concerns are considered as much as economic concerns (*ibid.*: 424).

This aspect of SUGC raises a host of questions related to the implementation of responses to cross-cutting issues such as food policy. There is no one size fits all answer to this challenge, however a growing body of research examining the implementation of municipal food policy mandates and comprehensive

(continued)

Box 1: (continued)

municipal food strategies is building a rich knowledge base of insights (Mendes 2008; Rocha and Lessa 2010; Mansfield and Mendes 2013; Hatfield 2013; MacRae and Donahue 2013). Factors commonly explored in these studies include the role of the institutional location of food policy processes, staff and budget support, the degree of integration into normative mechanisms, champions supporting a new policy area, overall management, and the extent of meaningful partnerships with non-governmental actors.

Overall, SUGC clearly offers a useful lens that encourages new approaches to emergent issues such as food policy. Taken in conjunction with other tools and conceptual frameworks, SUGC can enhance the understanding of the strengths and challenges of taking a holistic systems approach to urban food policy and planning that is coordinated and collaborative.

References

Appadurai A (2001) Deep democracy: urban governmentality and the horizon of politics Environ Urban 13(2): 23–44

Harvey D (1989) From managerialism to entrepreneurialism: the transformation in urban governance in late capitalism. Geografiska Annaler Series B Hum Geogr 71(1): 3–17

Hatfield M (2013) Food policy & programs: lessons harvested from an emerging field. Portland Bureau of Planning and Sustainability. http://www.portlandoregon.gov/bps/article/416389. Accessed 10 Nov 2015

Healey P (2002) On creating the 'City' as a collective resource. Urban Stud 39: 1777–1792

MacRae R, Donahue K (2013) Municipal food policy entrepreneurs: a preliminary analysis of how Canadian cities and regional districts are involved in food system change. http://tfpc.to/wordpress/wp-content/uploads/2013/05/Report-May30-FINAL.pdf. Accessed 10 Nov 2013

Mansfield B, Mendes W (2013) Municipal food strategies and integrated approaches to urban agriculture: exploring three cases from the global North. Int Plan Stud 18(1): 37–60.

Mendes W (2007) Negotiating a place for 'Sustainability' policies in municipal planning and governance: the role of scalar discourses and practices. Space Polity 11(1): 95–119.

Mendes W (2008) Implementing social and environmental policies in Cities: the case of food policy in Vancouver, Canada. Int J Urban Regi Res 32(4): 942–967.

O'Donovan D, Rimland Flower N (2013) The strategic plan is dead. Long live strategy. Stanf Soc Innov Rev. http://ssir.org/articles/entry/the_strategic_plan_is_dead._long_live_strategy. Accessed 9 Nov 2015.

Reynolds B (2009) Feeding a world City: the London food strategy. Int Plan Stud 14(4): 417–424.

Rocha C, Lessa I (2010) Urban governance for food security: the alternative food system in Belo Horizonte, Brazil. Int Plann Stud 14(4): 389–400.

While A, Jonas AEG, Gibbs D (2010) From sustainable development to carbon control: the eco-restructuring of the State and the politics of urban and regional development. Trans Inst Br Geogr 35(1): 76–93.

2009). New networks, such as the Sustainable Food Cities Network in the UK, and the 2015 Milan Urban Food Policy Pact, signed by over 100 cities around the world, can give cities a louder collective voice with regard to food policy. The local level could thus have more influence over higher levels of food system governance.

Conclusion: Mobilizing Methodological Tools and Drawing Up an New Research Agenda

This chapter has drawn attention to several important considerations for cities that seek to take responsibility for food. Firstly, it is helpful to identify pressing issues in the local context, as they can inform the overarching aims of food-related policy that resonate with top local government priorities. Secondly, the local context, including powers and capabilities that exist at the local level but are framed by multilevel governance, determines what can be done with respect to food and which levers can be utilized to achieve the aims. Thirdly, it is important to explore complementary levers and instruments across policy domains, as well as civil society and private sector contributions. There is a need for methodological and process tools that go beyond a check-box approach in order to assist cities in devising food policies that are appropriate and realistic within the local political economy. Failure to take the political economy into account is likely to impair the policy development and implementation process.

In addition to these lessons, this chapter has also identified two important issues concerning cities' food-related strategies. Firstly, questions have been raised about the appropriateness of cities seeking to (re)localise food provisioning, either as a specific aim or as a lever towards achieving other aims, if the city is not located within an agricultural basin and has historically forged longer-distance trade links. Secondly, there are questions over how far the local level can really contribute to systemic change, beyond just impacting food experiences within the urban setting and the immediate vicinity.

These issues each signal a possible new direction for the role and influence of cities within the food system. Regarding (re)localisation, there may be potential for cities to relocalize governance and control over supply chains instead of (or in addition to) territorial relocalization of food provisioning. This could be done through mutually beneficial partnerships or twinning with nearby or remote agricultural areas, investing in their production capacity, setting standards and offering a guaranteed market. There is a need for research into the conditions that would facilitate such innovation and any practical and ideological barriers that could arise. Test grounds are also required. These could include an initial focus on direct relationships to procure food for public canteens, or the extension and adaptation of current models to govern direct trading principles, such as Fairtrade and buyers' cooperatives.

As for cities' influence at higher policy levels, on-going research is needed on the potential of the combined efforts of many cities to strengthen their collective voice. In particular, this includes monitoring the impact of initiatives such as the 2015 Milan Urban Food Policy Pact on the international food policy dialogue. Another possible way to exert a more systemic influence could be to seek and obtain buy-in of actors or organizations that are influential in multilevel governance but un- or under-represented in local food policy. These actors and organizations include major food businesses, lobby groups and political figures. To this end, research is needed to gain insight into the barriers to engaging these actors, such as accepted framings and definitions of sustainability and food security, accepted norms and ways of working, and the governance structures to which they are bound.

References

Abelman J (2015) Cultivating the City: infrastructures of abundance in Urban Brazil. Urban Agriculture Magazine 29: 62–64

Andersson G, Nillson H (2012) Policy for sustainable development and food for the City of Malmo. In: Wiskerke JSC, Viljoen A (eds) Sustainable food planning: evolving theory and practice. Wageningen Academic Publishers, Wageningen, pp 181–188

ARF (2014) Déclaration de Rennes: Pour des systèmes alimentaires territorialisés. http://www.diplomatie.gouv.fr/fr/IMG/pdf/de_claration-finale_ARF_cle81ab8a.pdf. Accessed 7 Aug 2015

Barling D, Lang T, Caraher M (2002) Joined-up food policy? The trials of governance, public policy and the food system. Soc Policy Adm 36(6):556–574

Blay-Palmer A (2009) The genesis of urban food policy in Toronto. Int Plan Stud 14(4):417–424

Born B, Purcell M (2006) Avoiding the local trap: scale and food systems in planning research. J Plan Educ Res 26(2):195–207

Braine-Supkova M, Gaspard A (2015) City region food systems on the political agenda in France. Urban Agriculture Magazine 29: 21–23

Brand C (2015) Alimentation et métropolisation: problématique vitale oubliée. Doctoral thesis in Geography, Université Grenoble Alpes, p 656

Braudel F (1979) The wheels of commerce: civilization and capitalism: 15th–18th century, volume 2. Armand Colin, Paris

Burgan M, Winne M (2012) Doing food policy councils right: a guide to development and action. Mark Winne Associates, Santa Fe, p 63

Carey J (2013) Urban and community food strategies. The case of Bristol. Int Plan Stud 18(1):111–128

Cohen M, Garrett J (2010) The food Price crisis and urban food (in)security. Environ Urban 22(2):467–482

Council of Europe (1995) Recommendation N°R.(95)19 of the committee of ministers to member states on the implementation of the principle of subsidiarity. Adopted by the committee of ministers on 12 October 1995 at the 545th Meeting of the Ministers' deputies, Council of Europe, Brussels

Dahlberg K (1995) Food policy councils: the experience of five Cities and one County. In: Joint meeting of the Agriculture, food, and human values society and the society for the study of food and society. Tucson, Arizona, 11 June 1994, p 22

Derkzen P, Morgan K (2012) Food and the City: the challenge of urban food governance. In: Wiskerke JSC, Viljoen A (eds) Sustainable food planning: evolving theory and practice. Wageningen Academic Publishers, Wageningen, pp 61–66

Detroit Future City (2012) The land use element: the image of the City, Detroit Future City, 2012, Detroit, http://detroitworksproject.com/wp-content/uploads/2013/01/DFC_Plan_Land-Use.pdf. Accessed 19 June 2015

Di Iacovo F, Brunori G, Innocenti S (2013) Le strategie urbane: il piano del cibo. Agriregionieuropa 32(March):9

Dupuis E, Goodman D (2005) Should we go 'Home' to eat? Toward a reflexive politics of localism. J Rural Stud 21(3):359–371

European Union (2010) Consolidated versions of the treaty on European Union and the treaty on the functioning of the European Union (2010/C83/01). Off J Eur Union, 53 (30 March 2010)

FAO (2009) How to feed the world in 2050. FAO, Rome, 35 p

Forster T, Hussein K, Mattheisen E (2015) City region food systems: an inclusive and integrated approach to improving food systems and urban-rural linkages. Urban Agric Mag 29:8–11

Freudenberg N, Silver M (2013) Jobs for a healthier diet and a stronger economy: opportunities for creating new good food jobs in New York City. New York City Food Policy Center at Hunter College, New York, 44 p

GLA (2011) The Mayor's food strategy: healthy and sustainable food for London: an implementation plan, 2011–13. Greater London Authority, London, 34 p

GLA, CIEH (2012) Takeaways toolkit: tools, interventions and case studies to help local authorities develop a response to the health impacts of fast food takeaways. Greater London. Authority and Chartered Institute of Environmental Health, London, 67 p

Halliday J (2015) A new institutionalist analysis of local level food policy in England between 2012 and 2014. Doctoral thesis. Food Policy specialization, City University, London, 294 p

Harper A, Alkon A, Shattuck A, Holt-Giménez E, Lambrick F (2009) Food policy councils: lessons learned. Food First, Institute for Food and Development Policy, Oakland, 63 p

Hay C (2002) Political analysis: a critical introduction. Palgrave Macmillan, Basingstoke, 336 p

Jennings S, Cottee J, Curtis T, Miller S (2015) Food in an urbanised world, the role of City food systems in resilience and sustainable development. Prince of Wales' International Sustainability Unit, London, 80 p

Johnson B (2013) 2020 vision: the Greatest City on earth, ambitions for London. Greater London Authority, London, 84 p

Laidlaw J (2015) Food security – a Core component of a Leading City's transformation agenda. UN Global Compact Cities Programme, Melbourne, 8 p

Lang T, Heasman M (2015) Food Wars. Routledge, Abingdon, 310 p

Lang T, Barling D, Caraher M (2009) Food policy: integrating health, environment and society. Oxford University Press, Oxford, 313 p

Libman K (2015) Has New York city fallen into the local trap? Public Health 129(4):310–317

Libman K, Freudenberg N, Sanders D, Puoane T, Tsolekile L (2015) The role of urban food policy in preventing diet-related non-communicable diseases in Cape Town and New York. Public Health 129(4):327–335

London, 2012 (2009) For starters: food vision fo the London 2012 Olympic games and Paralympic games. London, p 42

Mah CL, Thang H (2013) Cultivating food connections: the Toronto food strategy and municipal deliberation on food. Int Plan Stud 18(1):96–110

Mansfield B, Mendes W (2013) Municipal food strategies and integrated approaches to urban agriculture: exploring three cases from the global north. Int Plan Stud 18(1):37–60

Marsden T (2013) Contemporary food systems: managing the capitalist conundrum of food security and sustainability. In: Murcott A, Belasco W, Jackson P (eds) The handbook of food research. Bloomsbury, London/New York, pp 135–147

Mendes W (2008) Implementing social and environmental policies in cities: the case of food policy in Vancouver, Canada. Int J Urban Reg Res 32(4):942–967

Morgan K (2009) Feeding the City: the challenge of urban food planning. Int Plan Stud 14(4):341–348

Morgan K (2014) Nourishing the city: the rise of the urban food question in the Global North. Urban Stud 52:1379–1394

Morgan K, Sonnino R (2008) The school food revolution: public food and the challenge of sustainable development. Earthscan, London, 231 p

Morgan K, Sonnino R (2010) The urban foodscape: world cities and the new food equation. Camb J Reg Econ Soc 3(2):209–224

Morgan K, Marsden T, Murdoch J (2006) Worlds of food: place, power and provenance in the food chain. Oxford University Press, Oxford, 225 p

Orlando G (2011) Sustainable food vs. unsustainable politics in the city of Palermo: the case of an organic farmers' market. City Soc 23(2):173–191

Oursense City Council (2013) Ourense vegetable gardens. http://www.sustainable-everyday-project.net/urbact-sustainable-food/2013/10/24/ourense-vegetable-gardens-2. Accessed 29 June 2015

Paturel D, Marajo-Petitzon E, Chiffoleau Y (2015) La précarité alimentaire des agriculteurs. Revue Pour 225:77–81

Perrin C, Soulard CT (2014) Vers une gouvernance alimentaire locale reliant ville et agriculture. Le cas de Perpignan. Géocarrefour 89(1–2-3):125–134

Pierre J, Peters BG (2003) Introduction: the role of public administration in governing. In: Peters G, Pierre J (eds) Handbook of public administration. Sage, London, 640 p

Pothukuchi K, Kaufman JL (1999) Placing the food system on the urban agenda: the role of municipal institutions in food systems planning. Agric Hum Values 16:213–224

Reynolds B (2009) Feeding a world CITY: the London food strategy. Int Plan Stud 14(4):417–424

Rocha C, Lessa I (2009) Urban governance for food security: the alternative food system in Belo Horizonte, Brazil. Int Plan Stud 14(4):389–400

Roitman S, Bifarello M (2010) Urban agriculture and social inclusion in Rosario Argentina. Inclusive Cities Observatory, London, 7 p

Schiff R (2008) The role of food policy councils in developing sustainable food systems. J Hung Environ Nutr 3(2–3):206–228

Sonnino R (2013) Local foodscapes: place and power in the agri-food system. Acta Agric Scand Sect B Soil Plant Sci 63(S1):2–7

Sonnino R (2014) The new geography of food security: exploring the potential of urban food strategies. Geogr J 182(2):190–200

Sonnino R, Spayde J (2014) The 'new frontier'? Urban strategies for food security and sustainability. In: Marsden T, Morley A (eds) Sustainable food systems: building a new paradigm. Routledge, Abingdon/New York, 256 p

Soulard CT, Banzo M, Perrin C, Valette E (2015) Urban strategies and practices for agriculture and food: six mediterranean case studies. In: Proceedings of the second international conference on agriculture in an urbanizing society, Rome, 14–17 September

Steel C (2008) Hungry City: how food shapes our lives. Chatto & Windus, London, 400 p

Steel C (2012) Sitopia – harnessing the power of food. In: Wiskerke JSC, Viljoen A (eds) Sustainable food planning: evolving theory and practice. Wageningen Academic Publishers, Wageningen, pp 37–46

Stoker G (2000) Urban political science and the challenge of urban governance. In: Pierre J (ed) Debating governance: authority, steering and democracy. Oxford University Press, Oxford, 272 p

Van der Schans JW (2015) Developing the Rotterdam City region food system: acting and thinking at the same time. Urban Agric Mag 29:14–17

Velasco M, De Vrieze A (2015) Unlocking La Paz. Urban Agric Mag 2(29):70–71

Warhurst P (2012) TED talk: how we can eat our landscapes. Video, http://www.ted.com/talks/view/lang/en//id/1538. Accessed 29 June 2015

Wegener J, Seasons M, Raine KD (2013) Shifting from vision to reality: perspectives on regional food policies and food system planning barriers at the local level. Can J Urban Res 22(1):93–112

Wekerle G (2004) Food justice movements: policy, planning, and networks. J Plan Educ Res 23(4):378–386

Wertheim-Heck S, Vellema S, Spaargaren G (2015) Food safety and urban food markets in Vietnam: the need for flexible and customized retail modernization policies. Food Policy 54:95–106

Winter M (2003) Embeddedness, the new food economy and defensive localism. J Rural Stud 19(1):23–32

Wiskerke J (2009) On places lost and places regained: reflections on the alternative food geography and sustainable regional development. Int Plan Stud 14(4):369–387

Wiskerke J, Viljoen A (2012) Sustainable urban food Povisioning: challenges for scientists, policy-makers, planners and designers. In: Wiskerke JSC, Viljoen A (eds) Sustainable food planning: evolving theory and practice. Wageningen Academic Publishers, Wageningen, pp 19–36

World Food Summit (1996) Rome declaration on World Food Security. Food and Agriculture Organization of the United Nations, Rome

Yeatman HR (2003) Food and nutrition policy at the local level: key factors that influence the policy development process. Crit Public Health 13(2):125–138

Theoretical Approaches for Effective Sustainable Urban Food Policymaking

**Julie Debru and Caroline Brand with Vanessa Armendáriz,
Stefano Armenia, Alberto Stanislao Atzori, Nevin Cohen,
and Paul James (Contributions)**

The emergence of food strategies and policies in many cities worldwide (chapter "Urbanization Issues Affecting Food System Sustainability; Nicolas Bricas") has prompted researchers to investigate associated building, support and assessment processes. In the light of the limits of industrialized food systems, these strategies and policies are driven by the need for a transition towards more sustainable food systems. This chapter looks at the conceptual frameworks used by researchers to assess urban food system sustainability.

Diversity and Complexity

The main problem currently facing researchers and public, private and community actors is the need to account for the complexity of the issue, including the extent of stakeholder involvement, policy areas and the scope of governance and action (Brand 2015). Until recently, the food issue has been handled on a sectoral basis at international, European and national levels. The dimensions of this issue have been slotted into separate 'silos', so it has only been dealt with regard to its agricultural, commercial, normative health security and, more recently, public health aspects (Brand 2015). In 2007, Guillaume Dhérissard and Dominique Viel underlined the risk of vulnerability and perverse effects linked to segmentation of the food issue (productivity, health security, ecology, marketing, etc.), which could lead to disruptive situations. They pointed out the need to consider food as a complex social phenomenon with a complete change of approach in favour of sustainable urban food systems and governance.

J. Debru · C. Brand (✉)
UNESCO Chair in World Food System, Montpellier SupAgro, Montpellier, France
e-mail: carolinebrand@hotmail.fr

© The Author(s) 2019
C. Brand et al. (eds.), *Designing Urban Food Policies*, Urban Agriculture,
https://doi.org/10.1007/978-3-030-13958-2_4

Moreover, the first report of the International Panel of Experts on Sustainable Food Systems (IPES-Food) called for a more holistic view of food systems (IPES-Food 2015). They proposed to look at food systems as a network of complex interactions between actors and system processes, and as a network of policies and regulatory frameworks. The complexity paradigm suggested by Edgar Morin highlights that a *whole* is not simply the sum of its parts—the complexity binds the parts to the *whole* and the parts to each other. This paradigm provides a structure for our thoughts on sustainable urban food systems. It is thus understood, for instance, that urban agriculture considerations cannot be limited to a nurturing approach, but should also include issues pertaining to education, social ties, aesthetics, biodiversity conservation, etc. (Duchemin et al. 2010; Duchemin 2013).

That said, the hardest part is to determine how to identify levers that could improve the sustainability of food systems based on a holistic view, and how to take the complexity, diversity and totality of sustainable urban food systems into account. This chapter successively outlines ways of dealing with this complexity.

Scientists are addressing this complexity by developing conceptual frameworks and representation models to encompass various aspects of sustainable urban food systems within a common vision. These analytical and insight-generating approaches are associated with forms of representation and single- or multi-disciplinary conceptual references (geography, political science, agronomy, economy, sociology, etc.). They offer different ways of identifying problems, finding solutions and enhancing the sustainability of urban food systems, while mobilizing practical tools for their assessment (phosphorus and nitrogen flows, food kilometres, food desert mapping, food footprint, carbon footprint, etc.).

There are several possible integrated ways of dealing with the issue of sustainable urban food systems, by focusing on: the food system, food sectors, social practices, policy areas or instruments, sustainability issues, spatial representations, etc.

These conceptual frameworks are not always explicitly geared towards promoting sustainable urban food policymaking. They are also used to gain insight into and describe urban food systems, in addition to building simulation models and developing approaches to assess the impacts of specific policies, projects or initiatives.

Through this diversity of approaches, a range of different solutions may be proposed for building more sustainable food systems. Combinations of approaches and tools may be able to effectively account for the complexity of the sustainable urban food issue—it is not necessary to build a widely applicable blanket framework integrating all approaches for this task. This notion of arrangements and combinations is common in many research fields that address the sustainability issue, including nutrition through studies on individual dietary diversity and its impacts, and economics via combined policy assessments (Esnouf et al. 2011). As diversity is now recognized as being a resilience factor, a range of combined approaches should be considered as a way to achieve more sustainable urban food systems.

In this chapter we thus present three types of approach that we feel are effective for drawing up sustainable urban food policies: systemic approaches that strive to incorporate sustainability issues in the food system analysis; approaches developed for analysing and building sustainable cities while addressing food issues; and finally a sustainable development approach to urban issues and food.

Food System Approaches

Systemic and Modelling Approaches

Systemic approaches have been used to understand and describe how food systems function (Rastoin and Ghersi 2010). When implemented by agrifood economists, these approaches are based on a functional view of food, describing a chain of operations and sectors (production, processing, distribution, consumption disposal/ recycling, regulation). The food system has thus been defined as:

> [...] an interdependent network of actors (businesses, financial institutions, public and private bodies) located in a given geographical area (region, State and plurinational area), while directly or indirectly participating in the creation of flows of goods and services geared towards fulfilling the food needs of several consumer groups locally or outside of the considered area (Rastoin and Ghersi 2010: 19).

The focus here is essentially on how the supply is organized.

Rationales have changed in favour of cyclical views, with the development of the circular economy concept, along with growing awareness that the waste produced by our food system is a resource. Analysis and assessment tools have thus been developed to gain greater insight into how territorial food system cycles work. For instance, on the basis of industrial ecology research, territorial ecology has given rise to the territorial metabolism concept, which makes it possible to get a snapshot of territorial food supplies: "[...] the analysis of material (raw), energy and substance flows, as well as the measurement of environmental footprints are concepts and methods that all contribute to this [territorial metabolism] characterization" (Barles 2014: 2).

Territories are seen as living organisms based on a 'lifecycle' rationale focused on inflows, 'digestion' and outflows (chapter "Urbanization Issues Affecting Food System Sustainability; Nicolas Bricas", Box 1.1). As Barles points out in her study on the urban metabolism of Paris, the global approach to flows is also interesting because it highlights the upstream and downstream dimensions (fertilization and waste management) of the food issue, which have yet to be sufficiently accounted for in thinking and initiatives:

> It reveals the need for new public policies, especially concerning waste management—to reduce construction material imports—and urban planning—to reduce their consumption. In addition, more research and the development of action plans to link urban and agricultural policies to improve the use of urban fertilizers and to favour local food supply are required (Barles 2009: 911).

Systemic approaches focus on the relationships between system components, their interactions and interconnections. For example, the model developed by

Vanessa Armendáriz and colleagues reveals the interdependence between the different components of food supply and distribution systems. This model has a system perspective and applies the system dynamics method to gain insight into food supply systems. Although developed to assess situations in developing countries, the model is otherwise especially interesting because it offers a representation that sheds light on the overall functioning of a given food system in interaction with others (habitat, movement, economy, technology, etc.). It shows the linkages and reciprocal influences between these systems, as well as cause-effect and carryover relationships between the different components. This facilitates identification of indirect causes or effects that may not be foreseen at the outset. It also serves as a research tool to simulate the impacts of policies targeting certain levers on the entire food system.

A Systems Approach to Urban Food Supply and Distribution Systems[1]

Vanessa Armendáriz, Stefano Armenia, Alberto Stanislao Atzori

This section presents a framework to gain a greater understanding of food supply and distribution systems (FSDS). The latter are described using system thinking (ST) and system dynamics (SD) approaches to highlight how the identification of FSDS feedback structures can guide policymaking to meet urban food needs.

A system perspective implies the presence of interconnected elements to fulfil a function or an objective over a given time period (Meadows 2008). Those elements can be physical or informative. When observing food systems, interrelationships may be detected between different elements involved in food production, supply, processing, distribution and consumption activities. A systemic and dynamic analysis can help gain insight into food system feedback loops and accumulation processes in urban environments. Accumulation processes determine changes in critical resources and drivers of food production and distribution, and are essential for assessing their sustainability.

What Are System Dynamics (SD)?

SD techniques may be used to assess a system structure, characterized by feedbacks among its parts. The system behaviour over time is the result of the system feedback structure, which can be qualitatively conceptualized through causal loop diagrams (CLD). Causal maps (or CLD) may be drafted to map feedback within and across

[1] The authors would like to thank the Food and Agriculture Organization of the United Nations (FAO/AGS – Rome, Italy) for providing valuable information and prior knowledge on the FAO FSDS Framework of Analysis, and for precious help in building the revised SD framework.

interacting subsystems. Two kinds of loops are studied on the basis of their characteristics:

1. Reinforcing: self-reinforcing loops. This implies that the system grows exponentially if these loops are dominant or the sole ones in the system.
2. Balancing (B): self-correcting loops that counteract change. Balancing loops seek equilibrium.

The system dynamics (SD) arise from the interacting complex network of these two kinds of loops (Sterman 2012). The SD can be analyzed through simulation after building a stock and flow diagram (SFD), also referred to as an SD quantitative model. The model formalization consists of describing, through an SFD, involving the presence of differential equations, how the system variables are interconnected and how the accumulation processes are determined by flow changes that alter the state of the system levels (or stocks).

System Dynamics (SD) Applied to Study Urban Food Systems

SD modeling is an iterative process to get a better understanding of the system (Ghaffarzadegan et al. 2011). As shown in Fig. 4.1, the modeling process requires the identification and definition of the problem. The overall system conceptualization results in a qualitative or quantitative model formalization, which often

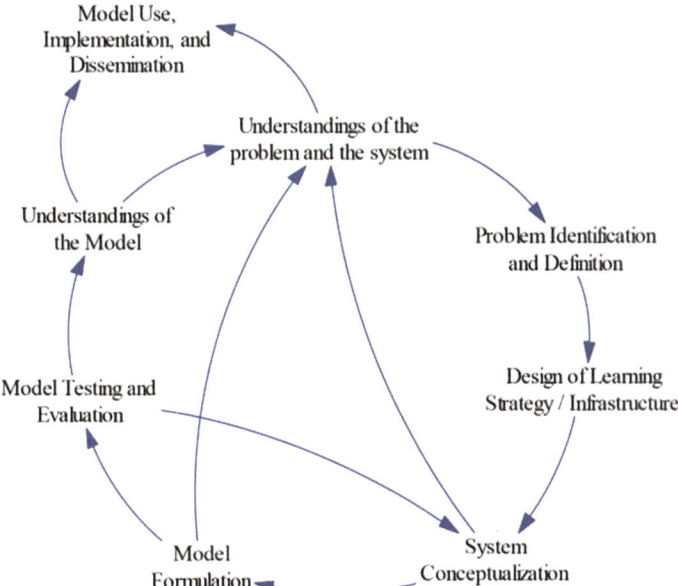

Fig. 4.1 Overview of the SD modeling process. (Zock 2004, adapted from Richardson and Pugh 1981)

improves our initial understanding of the system. The simulation model enables assessment of the model validity with empirical data, testing policy alternatives and gaining insight to increase the likelihood of performing a good policy analysis.

An example of the application of the SD method to study urban food systems is the FSDS framework setting model (Fig. 4.2) created on the basis of a detailed study leading to the publication of the Methodological and Operational Guide to Understand FSDSs (Aragrande and Argenti 2001) and complementary FAO documents (Argenti 1999a; b; Balbo et al. 2000; FAO 2000). The aim of this work was to get an overall understanding of urban food systems. This model was designed on the basis of the dynamics that generally prevail in developing and transitional countries. In the modelling phase, we decided to not include the characteristics of developed cities. This approach, in fact, required some simplification of the first detailed FSDS model (Armendáriz et al. 2015a) in order to capture—at an aggregated level, while maintaining its validity—the main system interactions, including non-food system issues.

In Fig. 4.2, blue arrows indicate a positive causal effect of an independent variable change on the dependent variable, i.e. when an independent variable increases or decreases, the dependent variable changes in the same direction. The red arrows indicate a negative causal effect among independent and dependent variables, i.e. when an independent variable increases the dependent variable decreases. The bold blue arrow represents the main question addressed in the FAO guide, i.e. "how can the urban food needs of a growing population be met?" The main feedback struc-

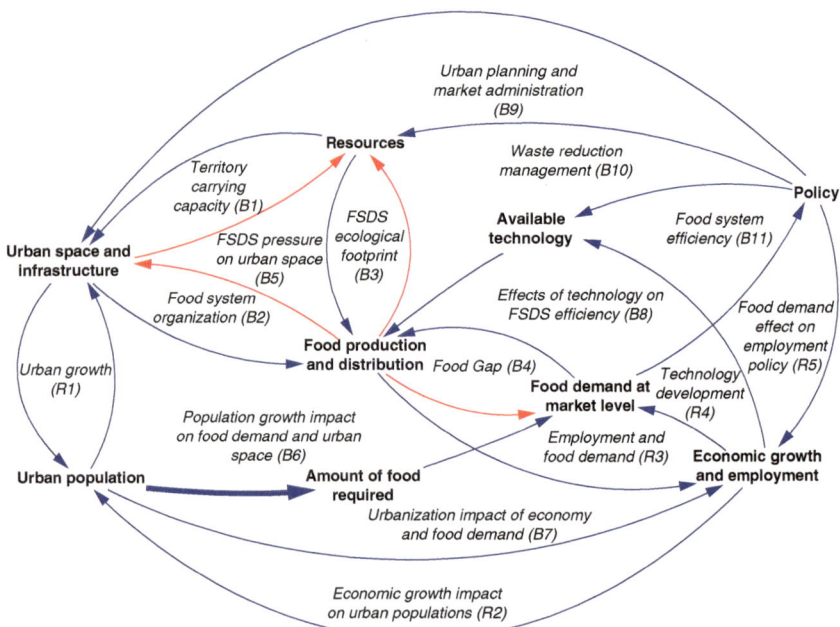

Fig. 4.2 FSDS framework setting model

tures are identified and named according to their function in the system: B1 Territory carrying capacity, B2 Food system organization, B3 Ecological footprint, B4 Food gap, B5 FSDS pressures on urban space, B6 Population growth impact on food demand and urban space, B7 Urbanization impact on economy and food demand, B8 Effects of technology on FSDS efficiency, B9 Urban planning and market administration, B10 Waste management reduction, B11 Food system efficiency, R1 Urban growth, R2 Economic growth impact on urban populations, R3 Employment and food demand, R4 Technology development, and R5 Food demand effect on employment policy.

The Urban growth (R1) system loop represents the feedback between the urban population and the urban space, which is a self-reinforcing relationship. Urban population growth increases the need for new infrastructure and urban space, meanwhile urban growth functions as a population attractor (World Bank 1999). The Economic growth impact on urban populations (R2) loop explains an additional self-reinforcing dynamic observed in developing and transition countries, i.e. economic growth in cities increases the rural population migration rate, in turn increasing the labour force for economic activities (UNDP 1999; Waibel and Schmidt 2000). Conversely, developed countries have different urban growth dynamics due to their different economic development patterns (Kastner et al. 2012).

The impact of urban growth may be noted in land-use changes, increased pollution and changes in non-renewable resource availability (Forrester 1970). This could be explained by the Territory carrying capacity (B1) of the system, which represents the maximum limit for continuing to produce or supply raw materials given the extent of resource depletion (Brenner 2014).

The relationship between urban space for markets and food logistics forms a balancing loop called the Food system organization (B2). Land devoted to urban activities (housing, services) reduces the space available for food markets and roads. Meanwhile, in the long-term, an increase in road coverage leads to an increase in urban agglomerations, thereafter increasing merchandise transport times and thus distribution costs (Aragrande and Argenti 2001). The FSDS Ecological footprint (B3) is explained by the environmental impacts of food production, processing, distribution and consumption activities involving renewable and non-renewable resource use, waste and pollution (Cordell et al. 2009; Ericksen 2008). The FSDS Pressures on urban space (B5) loop indicates that food system activities are only possible after urban space use and resource consumption issues have been addressed. The resource competition relationship between activities could lead to resource overconsumption, representing a risk for both the food system and the urban system.

Food system business revenues ensure economic growth, while the resulting employment opportunities provide income for the community, therefore increasing food demand at the market level. These relationships are represented by the Employment and food demand (R3) loop. The relationship between the food demand at the market level and food production and distribution activities is described by the Food gap (B4). Food production and distribution management deficiencies or even a resource crisis in the system could decrease the urban food supply and increase the food gap (Hanjra and Qureshi 2010; Pimentel and Pimentel 2007).

The Population growth impact on food demand and urban space (B6) represents the food demand linked with population growth, which boosts food production and distribution processes. However, population growth also requires an increase in urban space and infrastructure, which in turn leads to urban growth that attracts populations (Ehrlich and Holdren 1971; Waibel and Schmidt 2000). Another external balancing loop, i.e. the Urbanization impact on economic growth and food demand (B7), was discovered after assessing the population impact on economic growth. Employment increases income, thus increasing the food demand, while economic growth increases urban growth, which will boost the food production and distribution process, leading to competition with other industrial activities and housing for space and infrastructure (Allen and Sanglier 1978; Batty 2008; Pumain et al. 2015).

Indirect effects of the food market on economic growth and technological improvement are accounted for by the Technology development (**R4**) loop (Boserup et al. 1983). Enhancement of the food supply chain efficiency depends on the organization and technology levels applied to food production and processing, as explained by the Effects of technology on FSDS efficiency (B8) loop.

System Dynamics (SD) to Enhance Urban Food Policy

The results of a detailed analysis of FAO recommendations to policymakers in developing countries highlighted the target of improving food availability at the market level (Armendáriz et al. 2015a). Policy proposals were directly stimulated by variations in food demand. However, the FSDS framework (Fig. 4.2)—by identifying important interrelationships among urban elements and providing an aggregated view of the urban food system structure—revealed the following insight regarding the effectiveness of urban policies geared towards meeting actual food needs.

The urban food demand differs from the actual amount of food required in a city. The socioeconomic conditions determine the extent to which people are able to use their income to fulfil their food needs. Therefore a sound urban food policy should aim at reducing urban poverty levels. Economic growth and employment are needed to ensure access to food (Zezza and Tasciotti 2010; Von Braun 1995). The Food demand effect on employment policy (R5) loop explains how income is directly related to food consumption in terms of amount and quality. The socioeconomic status of the urban population is also closely related to health problems such as malnutrition and obesity. The impact of economic development on diet and health changes should be taken into account when planning interventions (McLaren 2007; Popkin 2001).

The Food system efficiency (B11) can be boosted by increasing the technology level used in food production and processing activities. Natural resource consumption is closely related to the efficiency of supply chains, consumer lifestyles and waste disposal processes. Environmental sustainability and urban metabolism indicators should be taken into account when designing food policies for rural and urban areas. Excessive use of natural resources due to the growth of human activities (i.e. in terms of both supply chains and consumption) could deplete system inputs and put the viability of the system at risk (Giampietro et al. 2013).

Waste reduction and management (B10) policies can improve the supply chain efficiency and reduce pollution. B10 and B11 structures represent balancing feedbacks that aim at increasing food supply by tackling urban inefficiencies and negative environmental impacts, while also reducing the urban food gap (Godfray et al. 2010; Parfitt et al. 2010).

Food policy and regulations (administrative protocols, regulations and laws) impact the quality of food to which communities have access. Different policy recommendations should be listed for the processing, distribution or consumption of different foods based on their core properties (high and low quality standards, prices, brands, food product lifecycles and availability of substitute products). For example, perishable and nonperishable foods have very different characteristics related to their lifecycle. Processing, distribution and consumption processes regarding these foods have constraints related to the available infrastructure, urban spatial organization, business logistics, consumer preferences, health risks, etc. (Beske et al. 2014). Apart from delays and side effects in some special food chains, there is a risk of causing urban food policy failures. For example, dairy products need a well developed refrigeration system, while an improvement in the storage capacity for dairy products would not be worthwhile if the distribution processes are limited by an absence of roads or by normal urban congestion and traffic, which would cause distribution delays.

Urban planning should be considered to optimize FSDS organization by setting up an effective food distribution infrastructure able to support both growing populations and the food gap (Pothukuchi and Kaufman 1999; Pothukuchi 2004; Born and Purcell 2006). The Urban planning and market administration (B9) loop considers key areas for policy intervention. These areas are related to urban spatial management and innovations, urban density and congestion, adequate roads and planned city allocation of formal and informal markets according to the urban population distribution and their socioeconomic characteristics. Loops B1, B2, R1, B5 highlight the indirect effects of such urban planning policies.

Preliminary outcomes from the studies on the application of the SD methodology to the understanding of FSDS (Armendáriz et al. 2015a, b, c) underlined that improving food system operations was not enough to cope with expected urban population growth, especially in developing countries. However, it turned out that urban population growth was not the main problem. The increase in urbanization processes is what actually causes the greatest pressure on food system support structures due to their impacts on land use patterns and their attraction for migrant populations and conurbations.

The Systems Approach for Food Policies in a Nutshell

A systemic view helps identify relevant issues beyond those related to food and which are often overlooked in urban food system analyses. The system dynamics (SD) method enables us to gain greater insight into interconnections between elements from which the structure of food systems can be analysed. Understanding how food systems work can support decision makers in identifying the actual problems and in formulating better policies to solve them, rather than just tackling the symptoms. Simulation exercises can even enable testing of different policies in a virtual environment. The example illustrated in this work has shown that it is essential to assess the environment in which FSDS are embedded. Understanding how the city is related to various physical or material issues (resources, technology, economy) is essential to gain insight into the origins of system pressures, while identifying possible intervention points for sustainability policies. The limitations of the FSDS model presented in Fig. 4.2 include: the focus on the FSDS structure and urban dynamics in developing and transition countries, especially megacity trends; the fact that the model was built on the basis of the FAO guide, with the aim of understanding FSDS in developing countries, and; the model's high level of abstraction, with the FSDS framework still being a qualitative approach to urban food systems. More comprehensive and robust policies are possible by integrating different methodologies in urban food system analyses, including agent-based modelling (ABM), structural network analysis (SNA), and geographical information systems (GIS).

Adaptive System

Systemic and modeling approaches provide an overview of the state of food systems, especially regarding the food supply organization and functioning mechanisms. However, as highlighted in the previous conclusions, these approaches must be combined with other tools to be able to gain insight into the full complexity of sustainable urban food systems.

For instance, it is hard to incorporate shifting uncertainties and dynamics related to stakeholders' practices into modelling approaches, but territorial ecology studies—at the crossroads between systemic and modelling analyses and more socially-oriented analyses on stakeholders (Barles 2010)—may help overcome this problem. Moreover, Debuisson (2014) combined a quantitative material and energy flow approach with a more qualitative approach that included role playing and modes of stakeholder interaction in the organization of these flows in order to analyse food and energy systems and understand how sustainable territorial dynamics become anchored. Approaches that stem from the complexity paradigm focus on 'adaptive systems', as opposed to 'deterministic systems' (end states) whose behaviour can be predicted (Cloutier 2013). In an adaptive system, the same stimulus can produce two different behaviours since they are dependent on the relationships between elements and not on the individual rationales of these elements. Hence, this approach can account for the fact that the same stimulus could trigger different reactions depending on multiple factors such as time, desire, humour, personal history, hunger, etc. This complex adaptive systems approach also helps us gain a more dynamic understanding of how urban food systems work.

This illustrates the advantage of combining the systemic approaches presented here with approaches that help gain insight into the practices and action capacities of system actors (who are linked with the sociopolitical, human, historical and cultural setting) as a way to grasp the complexity and sustainability of urban food systems.

Territorial Food Systems

A territorial view of food systems has more recently developed, particularly via the globally renowned city region food systems (CRFS) concept[2] and the territorial food system (TFS) concept[3] developed in France, both of which encompass the territory and agrifood sector concepts (Rastoin 2015). They are in keeping with literature that has emerged since the outset of the new millennium regarding the development of so-called 'alternative', 'regional' and 'local' food systems to better address sustainability challenges (Kneafsey 2010; Feenstra 1997, 2002; Feagan 2007; Hendrickson and Heffernan 2002).

These new concepts place the food system in a political, cultural, historical, agricultural and landscape setting. They contend that urban regions (a city associated with a more or less extensive supralocal area) have a key role to play in governance and in enhancing food system sustainability. These concepts reflect the hypothesis that a territorial approach could help solve some of the problems outlined in chapter "Urbanization Issues Affecting Food System Sustainability; Nicolas Bricas". They provide solutions for the issue of detachment (geographical, cognitive, economic) between urban and rural citizens, consumers and producers.

CRFS offer better integration of urban and rural issues regarding food supply through strengthened relationships between these two spaces (Jennings et al. 2015). Without simply advocating food autonomy or 'localism' as a blanket solution, territorialization of the food system concept has been put forward as a way to regain control of a system that eludes us in its global scope. The aim is to build a comple-

---

[2] A city region food system is defined as: "the complex network of actors, processes and relationships to do with food production, processing, marketing and consumption that exist in a given geographical region that includes a more or less concentrated urban centre and its surrounding periurban and rural hinterland; a regional landscape across which flows of people, goods and ecosystem services are managed." (Jennings et al. 2015)

[3] A territorial food system is defined as: "a consistent set of agrifood chains located in a geographical area of regional dimension. This concept focuses on maximizing the local integration of sectors as opposed to long globalized agrifood chains" (Rastoin 2015). In line with the mandate of the United Regions Organization (ORU Fogar), in the framework of the International Year of Family Farming, the Association of Regions of France published the Declaration of Rennes for Territorial Food Systems (TFS). This declaration underpins the position of the association in favour of promoting initiatives to entrench agrifood systems and their defense in local national and international public policies.

mentary alternative to the agroindustrial model that prevails with regard to food systems (Rastoin 2015).

These approaches mainly aim to encourage urban governments to account for food in their policymaking, in addition to the impacts of their policies beyond the territories they administer. They conceptually contribute to gaining greater insight into scaling issues by suggesting that multiscale food challenges should be taken into account locally.

Sustainable City Approaches

The territorial approach to food systems dovetails with other sustainable city planners' approaches. The food issue has, since the beginning of the new millennium, been placed back on the urban agenda of developers (Pothukuchi and Kaufman 2000). These approaches tackle urban food systems by taking spatial occupancy and organization or individuals and their practices into consideration—a viewpoint that complements that of the systemic approach.

Sustainable City Planning

The sustainable development target was promptly taken up by cities to such an extent that an 'urban shift' has been noted with regard to urban territorialization (Emelianoff 2007). Sustainable city concepts, indicators and models have been developing since the 1980s, in addition to the emblematic United Nations Agenda 21 action plans for sustainable development.

Sustainable city thinkers recognize the need for a change to a more sustainable and resilient urban system. Problems facing cities are not the specific problems identified in chapter "Urbanization Issues Affecting Food System Sustainability; Nicolas Bricas" but may be approached via the broader sustainable development concept, which is focused on concerns such as climate change mitigation, air, soil and water quality enhancement and rational resource management (water, energy, soil and biodiversity). These global issues are associated with specifically urban challenges such as urban growth, sanitary issues, water and energy supply, socio-economic and sociocultural issues, waste management and spatial planning. Urban architects, planners and developers thus design sustainable city models geared towards mainstreaming all urban sustainable development issues under one umbrella. These models are shaped by life science concepts, as well as economic, social and political science concepts and are based on cycles and resilience factors, e.g. the symbiotic city (Ranhagen and Groth 2012) and biophilic city (Beatley 2010) approaches. These sustainable city approaches overlap sustainable development issues with a variety of fields of urban action. For instance, the SymbioCity approach developed by Ulf Ranhagen and Klas Groth (2012) interlinks the following fields:

energy, architecture, water supply, waste management, industry and infrastructure, landscaping, urban transport and traffic, information and communications.

Food is not treated separately under these approaches and is at best one of many elements in the overall design of cities.

Urban Food Planning

Food is the focus of a targeted approach in the urban food planning field. Since the 2000s, a movement that brings together researchers and practitioners has been spreading in North America and northern Europe focused on integration of the food challenge in planning in association with the urbanization process. This field of research and action is structured on the basis of the noteworthy absence of food in the purview of planners. The terms used with regard to this absence highlight the sudden awareness of a missing piece in the thinking: "a puzzling omission" (APA 2007), "the dark side of urban dwelling?" (Viljoen and Wiskerke 2012), and "this intellectual lacuna" (Morgan 2015).

First, Pothukuchi and Kaufman (1999, 2000) showed that food is a key element in the functioning of the territories and as important as habitat and mobility which are the focus of urban planners' interventions. Secondly, the urban environment has been identified as an essential framework for showcasing faults and new food practices (Sonnino 2009; Morgan and Sonnino 2010; Morgan 2015). A 'new food equation'[4] is essential for the future development of areas undergoing an urbanization process (Morgan 2009; Morgan and Sonnino 2010).

This field is part of a new trend and mindset linked with the urbanization process and the territorial sustainability paradigm, where the role of actors involved in the food system as well as the planning of urban and metropolitan areas are reconsidered from a food perspective (Fig. 4.3).

Urban planning is a tool for building more sustainable and equitable food systems through a range of policy areas, while also reconsidering the status of food systems in production and spatial organization mechanisms (especially urban). Territorial configurations provide testing grounds for responses to major global issues in which the new food equation is nested.

This field offers an integrated vision of food systems whose governance includes civil society, private and public actors (Fig. 4.4).[5]

---

[4] The new food equation is based on the sudden increase in food prices in 2007–2008, which gave rise to new food security issues of global scope with regard to quantitative aspects of food supply, climate change, conflicts around arable lands and the urbanization process (Morgan 2009; Morgan and Sonnino 2010).

[5] In Figure 4.4, we have retained the different items concerning the food issue as presented in the schematic diagram of Johannes S.C. Wiskerke (2009) because they seem clearer than in the FoodLinks research report of Ana Moragues et al. (2013). However, we have kept the actor categories presented in the latter report because they seem more precise than those presented in Wiskerke (2009). We have not retained Weskerke's categorization and characterization of relationships

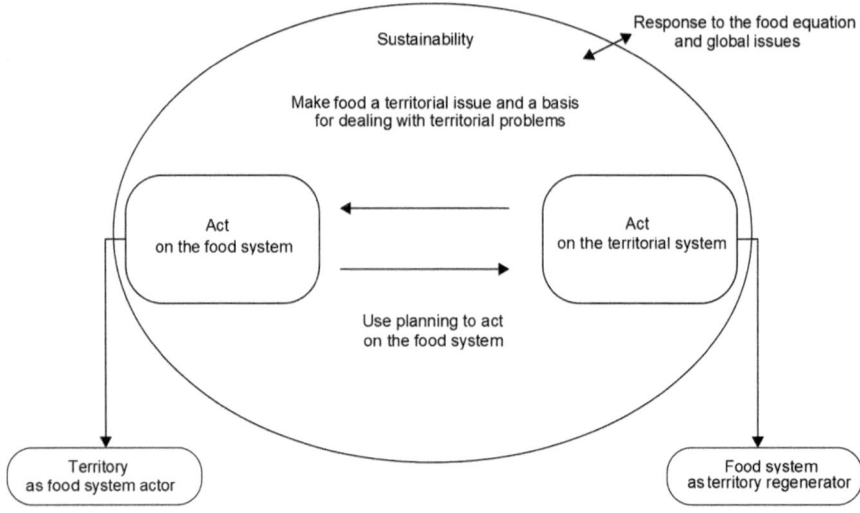

Fig. 4.3 Urban food planning – linking sustainability, the food system and the urban system. (Source: Brand 2015)

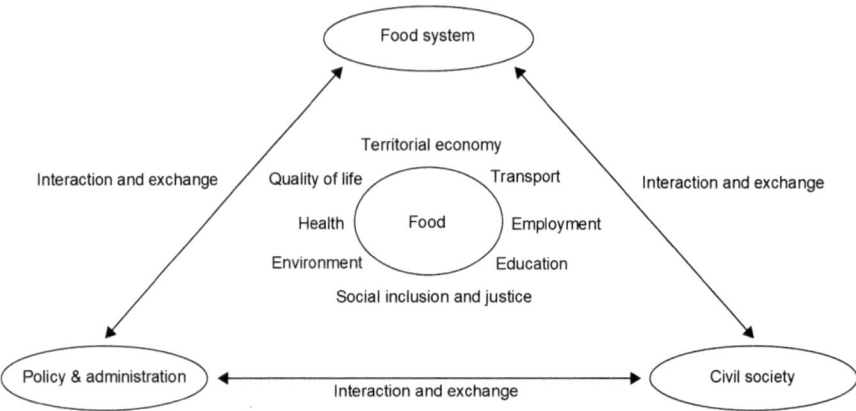

Fig. 4.4 Food governance described and considered in urban food planning. (Source: Brand 2015, from Wiskerke 2009: 376 and Moragues et al. 2013: 6)

between actors because they seem too restrictive of the actual and potential types of relationships between these actors with regard to the food issue (limited in the article to public food procurement, alternative food supply chains and urban food strategies). Moreover, in the 'urban food strategies' category, Weskerke's diagram indicates a relationship between government and civil society. We consider, however, that these strategies should also include economic actors in the 'market' actors sphere.

Urban planners, developers and architects must take the food issue—in all of its complexity—into account in sustainable city management and planning. Practical applications have been reported, such as the food sensitive urban planning and design framework (Donovan et al. 2011) in Victoria (Australia) and the Food Urbanism Initiative (Verzone 2012) in Lausanne (Switzerland), as well as the continuous productive urban landscape concept designed with regard to urban density (Viljoen et al. 2005). The aim is to link the food system, or associated elements such as agriculture, with urban planning.

The interaction between urban space and inhabitants' health in relation to food has been the focus of studies, giving rise to the foodscape concept, i.e. the food landscape—how the urban environment has an impact on food access or on the development of food-related diseases. At the crossroads of health planning and geography, researchers focus especially on the location and characterization of the type of food distribution outlets (Cummis and Macintyre 2002), e.g. studies carried out in the United States on the impact of the urban food environment and built environment on obesity (Raja et al. 2010). The authors showed that the proximity of a supermarket or grocery store to the consumer's home was correlated with women's body mass index (BMI). This concept has revealed new levers for local urban governments.

The organization of urban space has thus proven to be a significant determinant with regard to the food practices of city users (inhabitants, workers, tourists). The recent and ongoing rising awareness of urban planners, developers, architects and leaders concerning the food issue has prompted these actors to focus on the food system. This has fostered dialogue between their disciplines and other food system oriented disciplines (agronomy, management, economy, sociology, nutrition, etc.).

Approaches specifically concerned with the interaction between individual behaviours and the urban space complement these spatial organization oriented approaches. They consider other food system components, such as culture, food knowledge and individual consumer behaviours. The sociology of consumption, which is focused on individuals and their choices, along with other disciplines (e.g. environmental psychology) develop approaches that integrate factors that consciously drive individuals and factors that more unconsciously shape their habits. Behavioural theories that have until now fueled awareness-raising, information and education policies for individuals have been widely questioned (Lahlou 2005; Stø et al. 2008). Among these new approaches, the theory of practices provides an innovative framework, thus breaking away from behaviourism practices that are considered inefficient. Cohen and Ilieva (2015) have suggested using this innovative practice theory to provide cities with effective levers and facilitate the transition to a more sustainable urban food system.

Theory of Social Practices for a Sustainable Multidimensional Urban Food Approach

Nevin Cohen

Like all large-scale sociotechnical systems, the food system is so entrenched that it seems unchangeable. Yet it is composed of and shaped by everyday mundane and habitual social practices. Hence it can be modified by policies geared towards changing these practices (Shove et al. 2012; Watson 2012).

Ubiquitous everyday food practices have significant environmental and public health impacts, for instance cooking, refrigeration and dishwashing together account for 25% of household electricity consumption in the United States (Canning et al. 2010), while eating at a restaurant an extra day a week instead of at home can add a bit less than 2 kg a year to a person's weight. Such practices can thus generate a set of diet-related illnesses such as obesity (Todd et al. 2010). Conversely, these seemingly inconsequential food practices—which are often overlooked by food system planners because of their ordinariness—can turn out to be strategic levers to achieve broader social and environmental goals.

Role of Social Practices in Food Policy

Social practices are the everyday routines that people perform throughout their lives. They consist of *meanings* (beliefs, cultural norms and conventions), *materials* (tools, technologies, and financial resources), and *competencies* (know-how and skills) (Shove et al. 2012). Practices are *social* because they are shaped and reinforced by shared understandings of what is considered ordinary and appropriate ways of doing things. They are thus distinct from behaviours, which are actions based on the decisions of atomized individuals.

Practices are performed in the framework of interdependent sets of practices (Schatzki et al. 2001), so changes to one practice will therefore affect the entire set of practices.

Individuals are *carriers* of practices (Seyfang et al. 2010). They follow the rules and norms and, through their repeated performance of practices, they reproduce and normalize them and enrol others in the practices. According to Anthony Giddens (1984), their role as *carriers* is constrained by the structures and norms that arise concomitantly as they shape the practices. But individuals can also innovate and vary the way they perform practices, which are dynamic and can be reconfigured, leading to the emergence of new practices while others may disappear (Warde 2005). For example, conventional food canning practices have been supplanted by the freezing process. This is accompanied by a change in the meaning and skills associated with household food preparation practices and in the way of shopping.

Despite attention to environmental and social conditions and other upstream factors that contribute to obesity (Story et al. 2008; Freudenberg et al. 2015), policymakers have mainly focused on interventions geared towards changing individual behaviours rather than social practices (Warde 2014; Delormier et al. 2009). Behavioural interventions in the United States are generally designed to encourage healthier eating while reducing the incidence of obesity. They include mandatory disclosure of calorie information at fast-food restaurants to discourage excessive consumption. Moreover, financial incentives are proposed for new supermarkets to open in low-income communities in order to encourage fruit and vegetable consumption, or for individuals to buy healthier food at farmers markets (Cohen 2014).

The focus on individual behaviour is based on a framework that treats consumption as a function of the aggregated behavioural choices of individuals determined through individual preferences and rationally calculated assessments of risks and benefits (Halkier and Jensen 2011). Theories of behavioural change such as the planned behaviour theory and social marketing persist because of public health and policy conventions. They are in line with neoliberal ideology and relatively simple compared to more complex multi-sectoral approaches (Baum and Fisher 2014). Despite their popularity, behavioural approaches—which Shove (2010) criticizes as being an A-B-C (attitudes, behaviours, choices) theory of change—have not led to systemic changes in eating or to sustained reductions in the incidence of obesity (Cohn 2014).

In contrast, a social practice approach views that consumption models are embedded in the ways practices are structured. This approach does not overlook the effects of policies like regulations, consumer information or economic incentives to influence behaviour, but rather analyses and seeks to change unhealthy or unsustainable practices. Attention is paid to the meanings of a practice, the material dimensions and competencies associated with the practice. Focusing on the practices, how people perform them and what they use when carrying them out reveals the technologies and infrastuctures, cultural and social images and conventions, and knowledge that constitute these practices. The focus of this approach—rather than being about the ways individuals make food decisions—is on how their social practices (shopping, cooking, travel, dining with friends, taking a work break) structure their food practices.

Understanding the persistence and transformation of everyday food practices reveals opportunities for a transition towards healthier practices through policies that change the underlying meanings, materials and competencies. Consider for instance the practice of preparing food at home from scratch instead of eating less healthy fast food. Cooking should be a normal and socially desirable practice, not a burden that diminishes quality of life. Individuals need the requisite food storage and cooking equipment, in addition to time. Skills and know-how are also needed to prepare meals. If these three elements—meanings, materials and competencies— are not in place, the practice will not take hold and develop. Food practices must be viewed as part of a cluster of interdependent practices. A given practice may be changed (switching to cooking fresh products), along with associated practices (shopping on foot rather than by car) or how a practice is carried out may be modified (reducing the locations, e.g. classrooms or libraries, in which eating is considered acceptable). Hence, a social practice-based policy should also simultaneously consider a set of related practices associated with preparing meals at home (shopping, washing dishes, waste management, etc.).

Towards Strategic Management of Social Practices

Cities are uniquely positioned to engage in 'strategic practice management' (Cohen and Ilieva 2015), the process of stimulating new practices, re-establishing old ones, or changing the nature of existing ones. They can institute policies and programmes that could foster new meanings of practices, change the material and infrastructure conditions under which practices are performed, while altering competencies or disseminating new ones. Cities have many food practice transformation levers (Cohen and Ilieva 2015; 2016).

Cities—as centres of media and dense social networks—are able to support the creation of new meanings. Cities also run public school systems that offer strategic entry points to influence the practices of youth. Urban planning departments shape the spatial layout of practices by acting on the locations of fresh food retailers or authorising urban rooftop farming. Urban public health departments also influence the materials of urban food practices by setting nutritional standards or regulations for trans-fats, soda or salt. Moreover, cities can enhance the visibility of unconventional but healthy urban food practices.

Practice elements themselves can be sources of dynamism and challenge systems of practice. For example, making misshapen fruit culturally acceptable by serving it in public canteens can change buying practices and reduce food waste. Reviving school cooking instruction can change cooking competencies, stimulate home cooking and eventually increase fresh food purchases. Urban policymakers need to be attentive to practices with the greatest health and environmental effects. But they must also be mindful of weak signals and small changes not significant enough to transform entrenched sociotechnical systems. They can over time transform bundled practices, stabilize and become the new normal, eventually leading to broader change.

Moreover, food practices are connected to other sociotechnical systems like water management and transportation. Several sociotechnical systems can be adjusted at once by considering urban food systems as a complex of social practices (Cohen and Ilieva 2015). This provides an opportunity for cross-sectoral work. The practice approach induces organizational innovations, such as the creation of interdepartmental teams or multistakeholder task forces to work on more sustainable urban food systems, while suggesting that urban planning goals should be rethought, but not only in terms of land use, economic activities or infrastructure. Practices prompt reflection on the replication or reconfiguration of everyday practices that influence urban development. This generates new urban development expectations: a shift in focus to the elements that shape social practices; drawing up policies for sets of practices to enable withdrawal from conventional administrative silos; and a move away from models that assume an ability to predict behavioural changes based on interventions to boost individual awareness and information.

Approaches drawing on the theory of practices that are presented here have the advantage of not channelling attention towards economic, sociological and demographic determinants of individuals, as is often the case in consumer surveys. They can also be used to study the material, economic and social environment in which individual practices are carried out (Shove et al. 2012; Dubuisson-Quellier and Plessz 2013). Focusing more on the environment of practices so as to gain insight into the related behaviours broadens the prospects of policies geared towards influencing behaviours. They are intended to play a role in environmental change rather than boosting consumer awareness.

These approaches shed new light on policy instruments that local urban governments have at hand to enhance food systems and their sustainability. In dealing with urban management issues, food can serve as a tool to jointly meet various sectoral objectives borne by local urban governments in terms of health, wellbeing, ecology, economy, social equity, etc. Practice-based approaches offer a new framework to gain further insight into urban food sustainability determinants and to identify efficient levers for policymakers. They provide fresh opportunities for dialogue between researchers and policymakers.

Sustainable Development Applied to Urban and Food Issues

A final approach to review the complexity of food systems and formulate tailored policies should also be mentioned, i.e. sustainable development applied to urban and food issues. Sustainable development is an instrument for assessing the way societies work. Paul James and his colleagues developed an innovative sustainable development model applied to urban problems, and more recently to urban food issues, in a project carried out between 2007 and 2014 that was funded by the UN Global Compact Cities Programme in partnership with Metropolis and other international organizations. The Circles of Social Life approach aims to support cities, neighbourhoods and communities in the sustainable planning process. It combines qualitative and quantitative sustainability indicators, while drawing on a diverse range of actors with complementary expertise, to generate a holistic picture of the overall situation. This method has been applied particularly in Johannesburg, Melbourne, New Delhi, São Paulo and Tehran.

Towards a Holistic Understanding of Food Systems Via Circles of Social Life

Paul James

There is an Irish joke that provides an instructive way of thinking about useful beginnings and sustainable directions. The tale describes a disoriented traveller in rural Ireland asking directions of a local. The local begins to give detailed directions and each time falters, finally saying, "If it was meself that was going to Letterfrack, faith, I wouldn't start from here."[6] In addressing the question of sustainable urban food policies, we suggest the same thing, i.e. not beginning with the very issues that directly concern us most: food, urban settings, sustainability or policymaking. When focusing on the immediate areas of interest of specialists and professionals, the latter tend to conclude that their specific area is the most important starting point, while excluding all other possibilities.

Where then should the complex process of systematizing issues related to sustainable urban food policies be started? The usual place to begin is by approaching the food issue as an activity or economic value chain. But this starting point is restrictive with regard to urban sustainability, political power and cultural significance issues.

In Search of an Alternative Holistic Approach

One alternative—the circles of social life approach—begins with the question of the human condition, which encompasses food. Then how can we begin to depict that condition in a holistic way while identifying the food policy domains and subdomains? Most pressingly, how can we do so without being overwhelmed by the usual starting point of economics?

This is what food system analysts have started doing. For example, Geoff Tansey and Tony Worsley's work (1995) does this through a three-domain model. They begin with the biological domain (i.e. the living processes used to produce food), the economic and political domain (i.e. the power exercised over the food system) and the social and cultural domain (i.e. personal relationships, community values and cultural traditions that influence the way people use food) (ibid, p. 4). This blanket approach is much better than the commonly applied triple bottom line approach,[7] which includes three sustainable development spheres (Fig. 4.5).

The triple bottom line approach effectively considers economics as the master domain, with the environment as an externality that is costed against the economic aspects in accounting terms. The social domain pools all elements that do not fit the other two domains, including human rights, land rights, cultural identity, gender issues, etc. Hence, analysts' approach to the food system thus tends to better account for the social complexity, i.e. considering the economics without making it the starting point for all analyses. However, as soon as we start to look at the starting point for the work of Geoff Tansey and Tony Worsley, it quickly becomes clear that the prevailing complexity was not completely accounted for. Like the road to Letterfrack, the directions taken to develop a positive model for food system sustainability can soon get complicated. If one of their domains is biological, where in this approach

---

[6] The Hibbert Journal, vol. 22, 1924: p. 417.

[7] By the triple bottom line approach, the sustainable development concept is applied to the business sector.

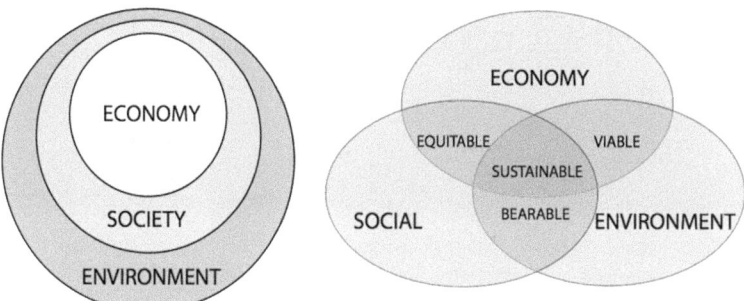

Fig. 4.5 Two depictions of the triple bottom line approach

are non-living and non-biological things which are so critical in food system functioning (e.g. nonorganic fertilizer and cadastral maps)? If their second domain is economic and political, does this mean that all economic questions are solely or primarily power issues? However, when striving to understand what it means when a peasant-farmer in the Andes sows seed in harrowed ground, it is essential (even if only for policy and analytical purposes) to separate questions of power—e.g. who controls the legal rights to seed reproduction of that seed—from economic questions such as what form of agricultural production frames seed sowing. Finally, if their third domain is cultural, where can the culture of capitalism, commodity fetishism and ideologies of growth be analysed? These too are cultural issues, but none of them begin with personal relations or cultural traditions. In short, their domains do not offer adequate generality and analytical coherence.

This underlines the importance of choosing the right methodological framework when conducting such studies. For circles of social life, circles of sustainability and circles of food, we worked with dozens of experts and local representatives worldwide and set up a lengthy dialogue process (James et al. 2015).[8] We sought to identify domains that would enable us to gain insight into the life of a traditional farmer respecting customs in the Andes and that of an agricultural futures trader in Paris. Four domains were first defined: economic, ecological, political and cultural. All of these were treated as social domains that could only be separated analytically, with the social domain always encompassed by and grounded upon the natural (Fig. 4.6).

Defining Social Domains

The ecological is defined as a social domain that focuses on practices, discourses and material expressions that occur at the crossroads between the social and natural realms. We recognize the distinction between these two realms in traditional (cosmological) and modern (scientific) understandings, with the natural being a context

---

[8] For further information on associated projects, see also: www.CirclesofSustainability.org and www.CirclesofFood.org

CIRCLES of SOCIAL LIFE
and beyond

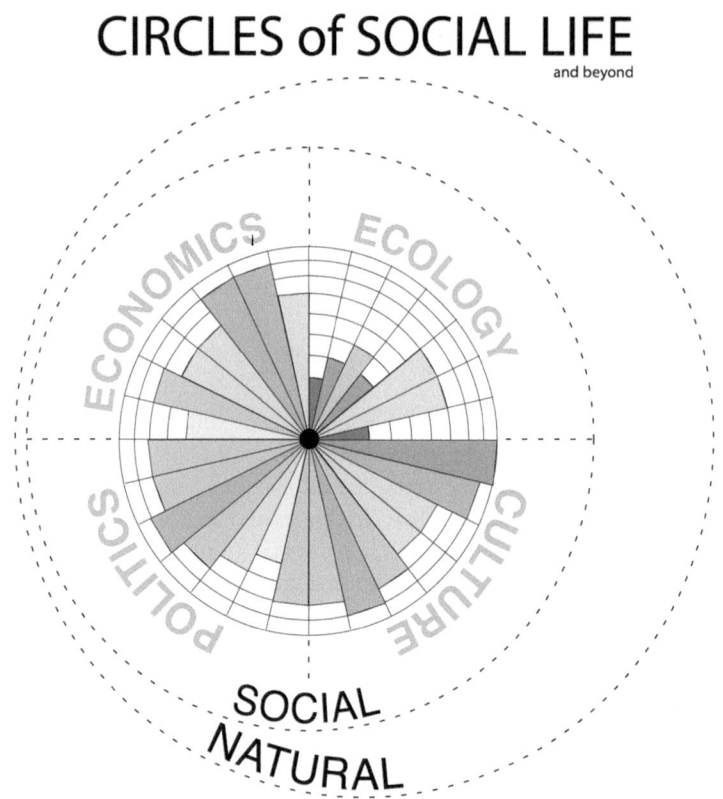

DOMAINS OF THE SOCIAL

ECONOMICS

Production & Resourcing
Exchange & Transfer
Accounting & Regulation
Consumption & Use
Labour & Welfare
Technology & Infrastructure
Wealth & Distribution

ECOLOGY

Materials & Energy
Water & Air
Flora & Fauna
Habitat & Settlements
Built-form & Transport
Embodiment & Sustenance
Emission & Waste

POLITICS

Organization & Governance
Law & Justice
Communication & Critique
Representation & Negotiation
Security & Accord
Dialogue & Reconciliation
Ethics & Accountability

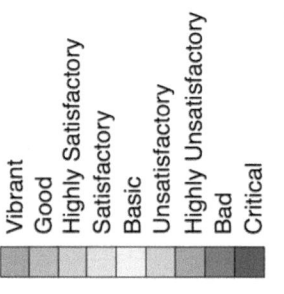

CULTURE

Identity & Engagement
Creativity & Recreation
Memory & Projection
Belief & Meaning
Gender & Generations
Enquiry & Learning
Wellbeing & Health

Fig. 4.6 Circles of social life understood in terms of four domains

of the social realm. But we enhance these two realms with human engagement with and within nature. This means that the ecological domain is focused on questions of interconnection between the social and natural, including the human impact on the environment and the place of humans in the environment. Foods may be sourced from nature, but as soon as they are grown, harvested or consumed they are no longer simply natural.

The economic is defined as a social domain that focuses on practices, discourses and material expressions associated with the production, use and management of resources. Here the resource concept is used in a broad sense, even in settings where resources are not instrumentalized or reduced to a means to achieve other ends. This allows us, for example, to compare different forms of food production and consumption. Although the economics domain was only differentiated as a named area of social life and deliberately practiced as a separate domain in the early modern period, by this definition it can be used in a broad range of places and time periods.

The political is defined as a social domain that focuses on practices and meanings associated with basic issues of social power as they pertain to the organization, authorization, legitimation and regulation of a social life held-in-common. This domain thus extends beyond the conventional sense of politics to include social relations in general. It bridges the public/private divide. The key related concept is a 'social life held-in-common' as many political issues bear directly on the sustainability of social life in general and food in particular. Consumption of a food in a certain way is political and thus concerns power.

The cultural is defined as a social domain that focuses on practices, discourses and material expressions, which express the continuities and discontinuities of the social meaning of a life held-in-common over time. Culture can be trivially defined as, 'how and why we do things around here'. The 'how' is how we practice materially, the 'why' emphasizes the meanings, the 'we' refers to the specificity of a life held-in-common, and 'around here' specifies the spatial and implicitly temporal particularity of culture. The culture concept had its beginnings in agriculture and cultivation, with subsidiary senses of 'honour with worship' of the word *cultura*, which in the sixteenth century was linked to the understanding of human growth and development (Williams 1976). This has obvious implications for food sustainability.

Seven subdomains are defined for each of these domains to provide tools for assessment, monitoring and evaluation. The approach strives to achieve sustainability and resilience through a combination of qualitative and quantitative indicators. It sets up a conceptual framework for investigating problems faced by communities and is intended to be applicable in very different neighbourhood, city and regional contexts.

Circles of Food in Practice

This method is currently being developed by Sustain: The Australian Food Network, with the aim of working with municipalities on their respective food systems.[9] Over the past 18 months, we have been working jointly to develop a food-profile process.

The first step in this process was to develop a food charter based on the four domains. The set of principles of the charter can be found online at: http://www.circlesoffood.org/principles.

Associated with these principles, the second step involved developing a set of questions for each of the four domains, their seven subdomains and seven aspects of each subdomain, for a total of 196 questions on food sustainability.

When taking the ecological domain as an example, the first subdomain is 'materials and energy' (Fig. 4.6). The following is a series of questions that were posed for the seven aspects of this food system subdomain:

1. Availability and abundance: how sustainable is the use of resources to produce food in the immediate region?[10]
2. Soil and fertility: to what extent are areas of arable land in the immediate region suitable for growing a variety of food produce?
3. Minerals and metals: how sustainable is the use of fabricated metals such as steel and aluminium in the food system across the broader region?[11]
4. Electricity and gas: is the electricity used in the various stages of the food system produced through ecologically appropriate and/or renewable means?[12]
5. Petroleum and biofuels: is the local food system overly dependent on fossil fuels?
6. Renewable energy and recyclable materials: does the local food system use recyclable materials?
7. Monitoring and reflection: does local monitoring of resource use result in the implementation of positive strategies relevant to the local food system?

The questionnaire is focused on the present period and the projection limits are the next 30 years, or one generation, according to the United Nations' definition of sustainable development. The idea is to meet current needs without compromising those of the next generation.

The series of questions are linked to indicators and a nine-level quality scale was drawn up. *Critical* is at the negative end of the spectrum, referring to an aspect of the food system requiring critical or urgent change now to ensure continuing basic viability over the next 30 years. *Vibrant* is at the positive end of the spectrum, refer-

---

[9] Project led by Kathy McConnel and Nick Rose.

[10] 'Immediate region' here means the area in question and its hinterlands. 'Material resources' includes all resources from water, food and energy to concrete and steel.

[11] 'Broader region' means within 3 h reach by land transport.

[12] Unless qualified by the adjective 'local', the 'food system' concept refers to the whole system upon which the local area depends—from local to global.

ring to an aspect of the food system that is currently active in reproducing vibrant social and environmental conditions enabling positive long-term advancement for the next generation and beyond. *Basic* is in the middle of the spectrum, referring to a quality that, at a pressure level equivalent to that of other levels, enables a basic balance to be achieved to meet the needs of the next generation over the coming period.

We suggest that each assessment ideally include 3–10 people from the targeted urban area who are specialists in different and complementary fields. In 2015, we conducted food assessments in three municipalities: Yarra Ranges with more than 100 people involved, including members of civil society, Whittlesea with 10 experts and Ararat with about 25 people representing regional organizations. In each case, when assessing the four social domains, we entered into the profile assessment process through discussion sessions on critical issues in the municipality. We asked each group to annotate their reasons for giving certain scores linked to these critical issues.

Food policies are currently being developed on each of these municipalities on the basis of the study findings. The future of this project will depend on how the method works in practice and contributes to developing a prosperous food system both in places where it is implemented and elsewhere.

Conclusion

This presentation of various approaches that are geared towards gaining insight into the complexity of the sustainable urban food issue highlights the promising potential of an integrated approach. The complexity of the topic warrants application of a combination of approaches. A systemic and cyclical view of food enables us to consider the activities, flows and sets of stakeholders involved and thus to understand the system as well as the underlying dynamics. The food system concept—recognized and used by scientific communities and food system stakeholders—thus helps to build a common vision of the complexity of urban food system sustainability.

The territorial approach to food systems in urban regions has the advantage of providing a framework for analysis and practice. It is an opportunity to facilitate political-scientific exchange and thus identify new prospects in the researcher-decision maker dialogue. It also provides a way to dovetail food sustainability and urban sustainability approaches and thus to focus maximum attention on a variety of determinants and levers.

Some of the theoretical approaches presented in this chapter already have practical applications through the development of local stakeholder support tools or measurement and assessment tools. Many food strategy action plans or documents outlined in the grey literature are structured according to a food system framework represented by a chain of activities (production, processing, distribution, consumption, waste management). They sometimes also include sectors like health and culture. To support the emergence of an urban food policy, practical application begins by an assessment of the situation, generally based on an analysis of socioeconomic activities (Carey 2011; Conley et al. 2011). Combinations of different analysis tools are implemented according to stakeholders' viewpoints and the local setting (territorial diagnosis from flow or lifecycle analyses or via qualitative methods, food desert mapping, food mile calculations or the ecological footprint, etc.).

However, urban food policymaking processes have timeframes of variable length and may be of different forms. The local history and political calendars are essential elements that preclude their emergence (chapters "History of Urban Food Policy in Europe, From the Ancient City to the Industrial City; Benoit Daviron, Coline Perrin, Christophe-Toussaint Soulard" and "Cities' Strategies for Sustainable Food and the Levers They Mobilize; Jess Halliday"). Some policies are promoted and identified at the international scale, while others are more discrete. Cities already have an impact on the food system through instruments that they use for management of land, school catering, social sector, economic activities, waste collection and disposal, etc. They sometimes have a targeted impact, e.g. on vulnerable populations, subsequently leading them to reflect on broader food issues. At other times, cities set up a crosscutting programme to coordinate sectoral initiatives to deal with food issues.

This diversity of processes questions the conceptual frameworks used to analyse and facilitate the formulation of urban food policies in all their forms and timeframes. The approaches outlined in this chapter inform us on two levels of understanding of this issue, i.e. the food system and its sustainability, and urban public action. These two levels are encompassed within broader frameworks for understanding sustainable development. These combined approaches can lead to the formulation of a monitoring and action framework at the crossroads between urban food system sustainability issues and levers implemented by local urban governments (chapter "Reconciling Sustainability Issues and Urban Policy Levers; Nicolas Bricas, Christophe-Toussaint Soulard, Clément Arnal").

References

Allen PM, Sanglier M (1978) Dynamic models of urban growth. J Soc Biol Struct 1(3):265–280

APA (2007) Policy guide on community and regional food planning. American Planning Association

Aragrande M, Argenti O (2001) Studying food supply and distribution systems to cities in developing countries and countries in transition. Methodological and Operational Guide. Food into Cities Collection, DT/36-01E, Rome, FAO

Argenti O (1999a) Urban food security and food marketing. A challenge to cities and local authorities. Food into cities collection, DT/40-99E. FAO, Rome

Argenti O. (Ed.) (1999b) Food into cities: selected papers. FAO Agricultural Services Bulletin, 132, Rome, FAO

Armendáriz V, Armenia S, Atzori A, (2015a) SD updates of FAO methodological guide to manage Food Supply and Distribution Systems (FSDS). Proceedings of the 33rd International System Dynamics Conference, Cambridge, MA, USA

Armendáriz V, Armenia S, Atzori A, Romano A (2015b) Analyzing food supply and distribution systems using complex systems methodologies. Proceedings 9th Igls-Forum on System Dynamics and Innovation in Food Networks, Innsbruck, Austria

Armendáriz V, Armenia S, Atzori A, (2015c) Understanding Food Supply and Distribution Systems (FSDS). First Mediterranean Conference on Food Supply and Distribution Systems, Rome, Italy

Balbo M, Visser C, Argenti O (2000) Food supply and distribution to cities in developing countries. A Guide for Urban Planners and Managers. Food into Cities Collection, DT/44-00E, Rome, FAO

Barles S (2009) Urban metabolism of Paris and its region. J Ind Ecol 13(6):898–913

Barles S (2010) Society, energy and materials: the contribution of urban metabolism studies to sustainable urban development issues. J Environ Plan Manag 53(4):439–455

Barles S (2014) L'écologie territoriale et les enjeux de la dématérialisation des sociétés: l'apport de l'analyse des flux de matières. Développement durable et territoires 5(1):1–19

Batty M (2008) Cities as complex systems: scaling, interactions, networks, dynamics and urban morphologies

Baum F, Fisher M (2014) Why behavioural health promotion endures despite its failure to reduce health inequities. Sociol Health Illness 36(2):213–225

Beatley T (2010) Biophilic cities: integrating nature into urban design and planning. Island Press, Washington, DC, p 208

Beske P, Land A, Seuring S (2014) Sustainable supply chain management practices and dynamic capabilities in the food industry: a critical analysis of the literature. Int J Prod Econ 152:131–143

Born B, Purcell M (2006) Avoiding the local trap scale and food systems in planning research. J Plan Edu Res 26(2):195–207

Boserup E, Makhoul N, Munn RE, Srinivasan TN, Robinson JA, Rocha C (1983) Population and technological change: a study of long-term trends. Int J Health Serv 13(1):15–31

Brand C (2015) Alimentation et métropolisation: problématique vitale oubliée. Doctoral thesis, Geography specialization, Université Grenoble Alpes, 656 p

Brenner N (2014) Urban governance – at what scale? LSE Cities, London School of Economics, UK

Canning P, Charles A, Huang S, Polenske KR, Waters A (2010) Energy use in the US Food System. ERR-94, US Department of Agriculture: Economics Research Service, March 2010

Carey J (2011) Who feeds Bristol? Toward a resilient food plan. Bristol City Council, Bristol

Cloutier JF (2013) Systèmes adaptatifs complexes – Historiques, propriétés et méthodes d'influences. In: Tuteur intelligent pour systèmes adaptatifs complexes. Unpublished dissertation

Cohen N (2014) Urban food systems strategies. In: Blanco H, Mazmanian D (eds) The Elgar companion to sustainable cities: strategies, methods and outlook. Edward Elgar Publishers, Cheltenham

Cohen N, Ilieva RT (2015) Transitioning the food system: a strategic practice management approach for cities. Environ Innov Soc Trans 17:1–19

Cohen N, Ilieva R (2016) Fooding the city: everyday food practices and the transition to sustainability. In: Finding space for productive cities: proceedings of the Sixth AESOP conference on sustainable food planning. Cambridge, Cambridge Scholars Publishing

Cohn S (2014) From health behaviours to health practices: an introduction. Sociol Health Illness 36(2):157–162

Conley B, Falk J, Hawes T, Jung YH, Kim GH, Maggiotto Jr T, Takahashi N, Wright T (2011) Room at the table, food system assessment of the Erie County. University of Buffalo Department of Urban and Regional Planning, Buffalo, NY. Prepared for the Erie County Department of Environment and Planning and the American Farmland Trust, 172 p

Cordell D, Drangert JO, White S (2009) The story of phosphorus: global food security and food for thought. Glob Environ Chang 19(2):292–305

Cummis S, Macintyre S (2002) A systematic study of an urban foodscape: the price and availability of food in greater Glasgow. Urban Stud 39(11):1704s–1709s

Debuisson M (2014) Les modes d'interaction pour une dynamique territoriale soutenable: un apport à l'écologie territoriale. Cas des systèmes d'approvisionnement alimentaire et énergétique. Doctoral thesis, Université de Technologie de Troyes, Sustainable Development specialization, 497 p

Delormier T, Frohlich KL, Potvin L (2009) Food and eating as social practice – understanding eating patterns as social phenomena and implications for public health. Sociology of Health and Illness 31(2):215–228

Dhérissard G, Viel D (2007) À problème complexe, solutions complexes. Économie et Humanisme (380):9–12

Donovan J, Larsen K, McWhinnie J (2011) Food-sensitive planning and urban design: a conceptual framework for achieving a sustainable and healthy food system. Melbourne, Report commissioned by the National Heart Foundation of Australia (Victorian Division)

Dubuisson-Quellier S, Plessz M (2013) La théorie des pratiques. Quels apports pour l'étude sociologique de la consommation? Sociologie 14(4):451–469

Duchemin É (ed) (2013) Agriculture urbaine: aménager et nourrir la ville. VertigO, Montreal, p 394

Duchemin É, Wegmuller F, Legault A-M (2010) Agriculture urbaine: un outil multidimensionnel pour le développement des quartiers. VertigO 10(2)

Ehrlich PR, Holdren JP (1971) Impact of population growth

Emelianoff C (2007) La ville durable: l'hypothèse d'un tournant urbanistique en Europe. L'Information géographique 3(71):48–65

Ericksen PJ (2008) Conceptualizing food systems for global environmental change research. Glob Environ Chang 18(1):234–245

Esnouf C, Russel M, Bricas N (2011) Pour une alimentation durable, Réflexion stratégique duA-LIne. Quæ, 286 p

FAO (2000) Food for the cities, food supply and distribution policies to reduce food insecurity: a briefing guide for Mayors, city executives and urban planners in developing countries and countries in transition. Food into cities collection, DT/43-00E. FAO, Rome

Feagan R (2007) The place of food: mapping out the 'local' in local food systems. Prog Hum Geogr 31(1):23–42

Feenstra G (1997) Local food systems and sustainable development. Am J Altern Agric 12(1):28–37

Feenstra G (2002) Creating space for sustainable food systems: lessons from the field. Agric Hum Values 19(2):99–106

Forrester JW (1970) Urban dynamics. IMR: Ind Manag Rev (pre-1986) 11(3):67

Freudenberg N, Franzosa E, Chisholm J, Libman K (2015) New approaches for moving upstream: how state and local health departments can transform practice to reduce health inequalities. Health Educ Behav 42(1 Suppl):46S–56S

Ghaffarzadegan N, Lyneis J, Richardson GP (2011) How small system dynamics models can help the public policy process. Syst Dyn Rev 27(1):22–44

Giampietro M, Mayumi K, Sorman AH (2013) Energy analysis for a sustainable future: multi-scale integrated analysis of societal and ecosystem metabolism. Routledge, London/New York

Giddens A (1984) The constitution of society: outline of the theory of structuration. Polity Press, Oxford

Godfray HCJ, Beddington JR, Crute IR, Haddad L, Lawrence D, Muir JF, Pretty J, Robinson S, Thomas SM, Toulmin C (2010) Food security: the challenge of feeding 9 billion people. Science 327(5967):812–818

Halkier B, Jensen I (2011) Methodological challenges in using practice theory in consumption research. Examples from a study on handling nutritional contestations of food consumption. J Consum Cult 11(1):101–123

Hanjra MA, Qureshi ME (2010) Global water crisis and future food security in an era of climate change. Food Policy 35(5):365–377

Hendrickson MK, Heffernan WD (2002) Opening spaces through relocalization: locating potential resistance in the weaknesses of the global food system. Sociol Rural 42(4):347–369

IPES Food (2015) The new science of sustainable food systems. Report 01

James P, Magee L, Scerri A, Steger M (2015) Urban sustainability in theory and practice: circles of sustainability. Routledge, London

Jennings S, Cottee J, Curtis T, Miller S (2015) Food in an urbanised world, The role of city region food systems in resilience and sustainable development. 92 p

Kastner T, Rivas MJI, Koch W, Nonhebel S (2012) Global changes in diets and the consequences for land requirements for food. Proc Nat Acad Sci 109(18):6868–6872

Kneafsey M (2010) The region in food – important or irrelevant? Camb J Reg Econ Soc 3(2):177–190

Lahlou S (2005) Peut-on changer les comportements alimentaires? Cahiers de nutrition et de diététique 40(2):91–96

McLaren L (2007) Socioeconomic status and obesity. Epidemiol Rev 29(1):29–48

Meadows DH (2008) Thinking in systems: a primer. Chelsea Green Publishing, White River Junction

Moragues A, Morgan K, Moschitz H, Neimane I, Nilsson H, Pinto M, Rohracher H, Ruiz R, Thuswald M, Tisenkopfs T, Halliday J (2013) Urban food strategies: the rough guide to sustainable food systems. Document developed in the framework of the FP7 project Foodlinks (GA No. 265287), 26 p

Morgan K (2009) Feeding the city: the challenge of urban food planning. Int Plan Stud 14(4):341–348

Morgan K (2015) Nourishing the city: the rise of the urban food question in the Global North. Urban Stud 52(8):1–16

Morgan K, Sonnino R (2010) The urban foodscape: world cities and the new food equation. Camb J Reg Econ Soc 3:209–225

Parfitt J, Barthel M, Macnaughton S (2010) Food waste within food supply chains: quantification and potential for change to 2050. Philos Trans Royal Soc London B, Biol Sci 365(1554):3065–3081

Pimentel D, Pimentel MH (eds) (2007) Food, energy, and society. CRC Press, Boca Raton

Popkin BM (2001) The nutrition transition and obesity in the developing world. J Nutr 131(3):871S–873S

Pothukuchi K (2004) Community food assessment a first step in planning for community food security. J Plan Edu Res 23(4):356–377

Pothukuchi K, Kaufman JL (1999) Placing the food system on the urban agenda: the role of municipal institutions in food systems planning. Agric Hum Values 16(2):213–224

Pothukuchi K, Kaufman JL (2000) The food system: a stranger to the planning field. J Am Plan Assoc 66(2):113–124

Pumain D, Swerts E, Cottineau C, Vacchiani-Marcuzzo C, Ignazzi CA, Bretagnolle A, Delisle F, Cura R, Lizzi L et Solène Baffi (2015) « Multilevel comparison of large urban systems »,

Cybergeo: Eur J Geogr [En ligne], Systèmes, Modélisation, Géostatistiques, document 706, mis en ligne le 17 janvier 2015

Raja S, Yin L, Roemmich J, Ma C, Epstein L, Yadav P, Ticoalu AB (2010) Food environment, built environment and women's BMI: evidence from Erie County, New York. J Plan Edu Res 29(4):444–460

Ranhagen U, Groth K (2012) The symbiocity approach: a conceptual framework for sustainable urban development. SKL International, Stockholm, 156 p

Rastoin J-L (2015) Systèmes alimentaires territorialisés en France. J Resolis 4

Rastoin J-L, Ghersi G (2010) Le système alimentaire mondial. Concepts et méthodes, analyses et dynamiques. Versailles, Quæ, 584 p

Richardson GP, Pugh AL III (1981) Introduction to system dynamics modeling with DYNAMO. Productivity Press, Cambridge, MA

Schatzki TR, Knorr-Cetina K, Von Savigny E (2001) The practice turn in contemporary theory. Psychology Press, Hove

Seyfang G, Haxeltine A, Hargreaves T, Longhurst N (2010) Energy and communities in transition – towards a new research agenda on agency and civil society in sustainability transitions. University of East Anglia, Centre for Social and Economic Research on the Global Environment (CSERGE), Working Paper EDM F, 10, 13

Shove E (2010) Beyond the ABC: climate change policy and theories of social change. Environ Plan A 42(6):1273–1285

Shove E, Pantzar M, Watson M (2012) The dynamics of social practice: everyday life and how it changes. Sage, London

Sonnino R (2009) Feeding the city: towards a new research and planning agenda. Int Plan Stud 14(4):425–435

Sterman J (2012) Sustaining sustainability: creating a systems science in a fragmented academy and polarized world. In: Weinstein M, Turner RE (eds) Sustainability science: the emerging paradigm and the urban environment. Springer, New York, pp 21–58

Stø E, Throne-Holst H, Strandbakken P et al (2008) Review: a multi-dimensional approach to the study of consumption in modern societies and the potential for radical sustainable changes. In: Tukker A, Charter M, Vezzoli C, Sto E, Munch Andersen M (eds) System innovation for sustainability 1: perspectives on radical changes to sustainable consumption and production. Greenleaf Publishing in association with GSE Research, Sheffield, pp 233–254

Story M, Kaphingst KM, Robinson-O'Brien R, Glanz K (2008) Creating healthy food and eating environments: policy and environmental approaches. Ann Rev Public Health 29:253–272

Tansey G, Worsley T (1995) The food system: a guide. Earthscan, Abingdon

The Hibbert Journal (1924) A quarterly review of religion, theology, and philosophy. 22

Todd, JE, Mancino L, Lin B-H, (2010) The impact of food away from home on adult diet quality. ERR-90, U.S. Department of Agriculture, Economic Research Service

UNDP (1999) Human development report. Oxford University Press, Oxford

Verzone C (2012) The food urbanism initiative. In: Viljoen A, Wiskerke JSC (eds) Sustainable food planning. Evolving theory and practice. Wageningen Academic Publishers, Wageningen, 598 p: pp 517–531

Viljoen A, Wiskerke JS (2012) Sustainable food planning: evolving theory and practice. Wageningen Academic Publishers, Wageningen, p 608

Viljoen A, Bohn K, Howe J (2005) Continuous productive urban landscapes. Designing urban agriculture for sustainable cities. Architectural Press, Oxford, p 304

Von Braun J (Ed.) (1995) Employment for poverty reduction and food security. International Food Policy Research Institute

Waibel H, Schmidt E (2000) Feeding Asian cities: food production and processing issues. FAO regional seminar Feeding Asian Cities, Bangkok

Warde A (2005) Consumption and theories of practice. J Consum Cult 5(2):131–153

Warde A (2014) After taste: culture, consumption and theories of practice. J Consum Cult 14(3):279–303

Watson M (2012) How theories of practice can inform transition to a decarbonised transport system. J Trans Geogr:488–496, http://www.sciencedirect.com/science/article/pii/S0966692312000889

Williams R (1976) Keywords: a vocabulary of culture and society. Fontana and Croom Helm, Glasgow

Wiskerke JSC (2009) On places lost and places regained: reflections on the alternative food geography and sustainable regional development. Int Plan Stud 14(4):369–387

World Bank (1999) Entering the 21st century. World development report. Oxford University, Press, Oxford, 300 p

Zezza A, Tasciotti L (2010) Urban agriculture, poverty, and food security: empirical evidence from a sample of developing countries. Food Policy 35(4):265–273

Zock A (2004) A critical review of the use of systems dynamics for organizational consultation projects. Deustche Lufthansa AG, Future European Operations

Reconciling Sustainability Issues and Urban Policy Levers

Nicolas Bricas, Christophe-Toussaint Soulard, and Clément Arnal

The development of urban food policies is not based on unified approaches. This policymaking process is relatively recent and there is still insufficient hindsight for the development of approaches that could encompass the complexity of food systems. In the previous chapters we have seen that many initiatives are under way around the world, taking very different approaches. Hence, research needs to focus on how to design such policies, and tailored methods are beginning to be formalized. As was shown in chapter "Theoretical Approaches for Effective Sustainable Urban Food Policymaking", some methods are based on the food supply system or value chain, and look at urban systems through that prism, whereas others build on a certain conception of the urban system and attempt to integrate agricultural and food considerations into urban sustainability scenarios. In either case, the knowledge that must be leveraged is multifarious and difficult to formalize. So great are the resulting uncertainties that the more general question arises as to the value of expertise in urban food policy making—a question already current with respect to environmental policies (Lascoumes 2010). While the players must clearly rely on evidence in order to decide and take action, they are confronted with a bewildering array of areas of endeavour, given the need to reconcile complex urban and food systems with sustainability issues.

In practice, we see that stakeholders are often not fully aware of that complexity when taking action or measures and are hence unable to predict all of the impacts on the various dimensions of sustainability. As the effects of their actions cannot be accurately foretold, they follow their intuition. Failures and partial misfires may result, but also successes and cases of systemic leverage. As this book has pointed

N. Bricas (✉)
UMR MOÏSA, Cirad, MONTPELLIER CEDEX 5, France
e-mail: nicolas.bricas@cirad.fr

C.-T. Soulard · C. Arnal
UMR Innovation, Montpellier University, INRA, CIRAD, Montpellier Supagro, Montpellier, France

© The Author(s) 2019
C. Brand et al. (eds.), *Designing Urban Food Policies*, Urban Agriculture, https://doi.org/10.1007/978-3-030-13958-2_5

out, city dwellers take many different initiatives, individually or in more or less formal organizations, seeking to overcome the constraints and limits of urban life.

To pay more heed to these initiatives, as is proposed by advocates of social practice theory, is not without political consequences. That theory suggests that public policies aimed at changing the impact of consumption on health or the environment should target practices and the environment that determines them rather than individuals and their knowledge and attitudes (Warde 2005). Similarly, it holds that the cumulative effect of changes in customary practices will bring systemic and institutional changes (Shove et al. 2012). In the urban food system context, as shown in chapter "Theoretical Approaches for Effective Sustainable Urban Food Policymaking", social practice theory would dictate that urban food strategies should be developed and consolidated synergistically, through a multiplicity of micro-initiatives (Cohen and Illieva 2015). Policies are no longer to be developed in isolation, without input from the citizens engaged in these initiatives, but rather based on citizens' practices and the transformative effects thereof.

Keeping in mind that desired manner of designing urban food policies, in this chapter we propose a conceptual framework under which cities' public actors will be better able to devise, manage and evaluate their food policies, or in other words to target and synergize the actions they take and assess their outcomes. These public actors, which we refer to as local urban governments (LUGs), are of course not the only stakeholders in the food system. They work alongside business and civil society, but we conceptualize their role as that of an enabler within the urban food system, able to spur these other actors to action or synergize them. The idea then is not to consider these local public actors in isolation, but to provide them with the tools they need, firstly, to enhance actions for which they could provide impetus via their expertise, and secondly to enhance their intermediary role vis-à-vis the other actors in urban food governance: the market and civil society (Viljoen and Wiskerke 2012). From that point of view, the food system sustainability issue is not solely concerned with the nature and effects of the actions taken, but also with the relevant policies' modes of governance (Brunori 2015). That duality is specifically illustrated in the account of the development of an urban food policy presented in chapter "Putting Food on the Regional Policy Agenda in Montpellier, France".

While chapter "Urbanization Issues Affecting Food System Sustainability" set out an analytical framework for urban food system sustainability, what we propose to explore here is an area that has been little studied in the literature: the means of action and leverage available to LUGs. Most approaches to the development of urban food policies call for a preliminary objective assessment of the urban food system to be addressed and identification of the problems to be solved. But such approaches seldom take the realistic scope of action of LUGs into account. Leverage, however, is another frequent consideration in their endeavours. The first part of this chapter deals with the various forms such leverage can take. Sustainability issues can be jointly plotted for comparative purposes on a double-entry matrix. Each lever being used can thus be assessed in terms of food system sustainability issues or, moving to the other side of the matrix, the different levers available to act on a given problem can be examined.

Urban food policies can then be developed through a back-and-forth between the identified problems and possible levers. The different ways of devising urban food policies will be analysed in the second part of this chapter.

Local Urban Governments' Scope of Action

Cities' renewed interest in the food issue is being expressed in the dual context of globalization and decentralization. LUGs are seeing a change in their jurisdiction and an increase in their powers, as evidenced for example by the implementation of territorial food projects (PATs) in France or Local Food Promotion Programs (LFPPs) in the United States. However, the degree to which State powers are decentralized to more local levels is very variable by country and often evolving in both respects, jurisdiction and scope of authority. In identifying the levers available to LUGs we do not mean to imply that all can always be deployed by all. Moreover, the typology proposed in this chapter is intended as a basis for reflection and needs to be adapted to each institutional and political situation.

Definition of Levers

In order to explore the levers available to LUGs, we propose to begin with the capacity for action that States endow them with: economic development, social, education or health policy, etc. To identify these levers, we looked at the internal organization of LUGs (administrative divisions, etc.) using the organization charts of a sample of world cities and conurbations for which that information was available.[1]

To be clear, we must differentiate these levers from the instruments, procedures or means available to an LUG to wield a lever. A 'lever' is defined as an LUG purview or area of activity. Spatial planning would be one example: an area of activity for which one or more initial objectives have been defined, for example the people's wellbeing, inequality reduction, or economic development. To achieve these objectives, the LUG has instruments of its own, such as its local urban plans (Plan Local d'Urbanisme, PLU), territorial coherence planning (Schéma de Cohérence Territoriale, SCoT), etc., in the case of France, and its own investment capacity. For a PLU or a SCoT, these instruments are based on various procedures: assessment, initial environmental state, sustainable planning and development plan (Plan d'Aménagement et de Développement Durable, PADD), orientation and objectives

---

[1] Cities and conurbations in France (Lyon, Lille, Bordeaux, Nantes, Montpellier, Annecy, Bourg-en-Bresse, Romans-sur-Isère, Alençon, Angoulême), in North America (Toronto, Quebec City, Montreal, New York City), in Europe (Lausanne, Geneva, Brussels, Liège, Bristol, London, Munich), and in Africa (Niamey, Dakar, Marrakesh, Tunis).

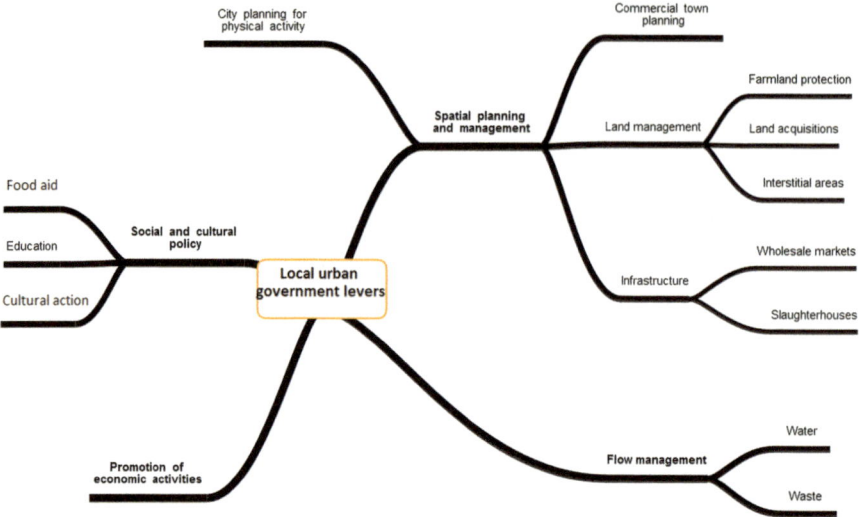

Fig. 1 Levers that local urban governments can deploy in support of agricultural and food policies

document (DOO), regulations, etc. Finally, LUGs have the human, technical or financial resources available to the units responsible for a lever or the use of an instrument. A lever—seen as an area of activity—thus possesses its own instruments, tools, procedures and means.

At an international conference that we held in Montpellier,[2] bringing together numerous Latin American, African and Asian cities that had implemented agricultural and/or food policies, we identified various levers used by them to ensure food security and improve popular nutrition or, more generally, to improve food system sustainability. Following that meeting we held a workshop with urban food systems experts at which we developed a typology of these levers and past and current objectives assigned to them.[3] As we discovered, those objectives have changed over time as new concerns emerge.

The main levers identified during this workshop are charted in Fig. 1 and described in the following paragraphs.

---

[2]An international meeting on 'Urban Food Policies; markets, catering, rural/urban connexions' organized by the UNESCO Chair on World Food Systems, CIRAD, Agropolis Fondation, FAO, the French Development Agency, RUAF Foundation, ORU-Fogar, the Agence Universitaire de la Francophonie, the International Sustainability Unit and the Foundation for the Progress of Humankind, held from 16 to 18 November 2015 in Montpellier (France).

[3]Workshop participants on 14 December 2015 were: Clément Arnal (consultant), Nicolas Bricas (CIRAD, UMR MOISA), Damien Conaré (Chaire UNESCO AdM), Benoit Daviron (CIRAD, UMR MOISA), Julie Debru (UNESCO Chair), Florence Egal (consultant, Milan Pact), Anna Faucher (IUFN), Paule Moustier (CIRAD, UMR MOISA), Dominique Paturel (INRA, UMR Innovation), Coline Perrin (INRA, UMR Innovation), Christophe Soulard (INRA, UMR Innovation).

Spatial Planning and Management

Cities are not exclusively built environments. They also include natural areas (forests, rivers, bodies of water and banks, etc.) as well as farmed or fallow land, in particular on the outskirts, whose future uses—maintenance or conversion—may be decided by LUGs. But built-up areas are also amenable to planning of alternate uses of 'free' areas or land uses related to urban projects: housing, infrastructure, shops, activities, etc.

Agricultural land needs to be protected from soil degradation and the leapfrog development associated with urban sprawl in order to maintain local agricultural production or create recreational or kitchen gardening areas, in particular for vulnerable populations. Several environmental objectives are fostered by these preservation efforts: biodiversity maintenance (green corridors), climate risk management (city cooling, buffer zones in case of flooding), greening of the city and creation of recreational areas and, lately, promotion of agricultural production for local food supply or even job creation. This farmland preservation may be done through regulation (see planning documents such as the local master plan or, in France, the territorial consistency plans (SCoTs), or farmland protection measures creating protected agricultural zones (ZAPs) and/or buffer zones around farmed and natural periurban land) and/or land acquisitions or rights of first refusal.

Some municipalities may acquire land well beyond their territorial limits, and in some cases have done so since ancient times. Lausanne, for example, owns eight farmsteads and some one hundred individual lots totalling about 900 ha. While all of this land was formerly leased out and worked independently of the city, Lausanne's new food policy has sought to take advantage of the land base to supply school canteens and develop educational farms as green technology showcases. The city's new agricultural policy also helps forge closer relations between town and country (Jarrige 2013).

On a finer scale, town planning regulations play on habitat density and the size and use of private plots. The choices made then determine the possibilities of cultivated gardens and domestic or shared orchards and hence food self-production. Similarly, marginal areas—fallow land, public gardens, schoolyards, sidewalks, vacant lots, building roofs—can be used for gardening, fruit production and animal husbandry (bees, hens), as a means of encouraging urban agriculture. Among the possible goals are: to provide access to sources of additional food or income under social policies, to create leisure areas or meeting places through gardening, to create places where people—and in particular children—can be introduced and sensitized to agriculture, to enhance management of the environment and livelihoods, and to promote health-giving physical activity. On this last point, designing town planning to facilitate physical activity can significantly improve people's nutritional status, particularly those who consume excess calories, through the provision of cycle paths, fitness trails and sports areas (Yin et al. 2013; Epstein et al. 2012; Raja et al. 2010).

When it comes to commercial town planning, LUGs can to a great degree dictate the location of food shops, markets and eateries. The food landscape is greatly affected by the establishment of malls on the outskirts or, conversely, the development of neighbourhood retail marketplaces—often for reasons of sanitation—and permanent or floating eateries (e.g. food trucks), all of which also makes a difference to how far households need to go for food supplies and to whether certain products are consumed as they become more or less accessible. Where no policy exists on food shop location, there is a danger that 'food deserts' will appear where it becomes difficult or expensive to easily find quality food, in particular fruit and vegetables. The reorientation of commercial town planning to prevent food deserts has become a concern for very many North American cities.

Flow Management

Substantial flows of food and water are needed to supply cities, which export their production and send their waste the other way. Everything that is consumed, recycled and discarded is affected by the management of these flows.

Wholesale market management, though initially concerned with the logistics of the flow of food into cities, can serve to connect or reconnect them to local farmers. City administrations that manage these wholesale markets can set aside specific direct sales areas to which small-scale producers, not just wholesale dealers, will have access. That is one priority for these markets, given the development of retail logistics that tend to reduce the number of reference products marketed and hence product diversity. Where the wholesale markets are sited, how producers access them, and how they interface with retail stores or markets are all determining factors for the conurbation's relationship with its hinterland. Proactive actions can encourage certain forms of agriculture by opening up the urban market to farms too small for the usual retail channels or by setting aside areas reserved for organic farmers.

The creation and management of municipal slaughterhouses on the outskirts of town generally serves a public health purpose: to eliminate the risks involved in having livestock within densely populated areas and to monitor the health of slaughter animals. In addition, they give nearby farms access to the urban market and favour the development of businesses to process meat, hides, horns and other by-products. Slaughterhouses also give structure to the economic organization of livestock industries.

City administrations often have jurisdiction over the provision of drinking water. In developing countries, that has a direct effect on nutrition in that it prevents infectious digestive diseases that are largely responsible for diarrhoea and hence childhood malnutrition. In industrialized countries where groundwater is polluted by agriculture chemicals, LUGs can acquire land within watersheds (e.g. Munich, New York) to ensure that it is not polluted, or give incentives to farmers operating within the catchment area to switch to organic farming in exchange for guaranteed outlets for the food they produce, to supply school cafeterias or municipal establishments (e.g. Munich, Lons de Saunier) (Le Strat 2008; Hellec et al. 2013). In that way cities can play an incentive role with respect to agricultural production conditions in areas outside their direct span of control.

Similarly, some LUGs contract with producers even in relatively remote rural areas to ensure quality food supplies. In exchange for a commitment to selling their produce to the city, LUG services can then advise them and even promote investment in the rural areas concerned. The municipality of Hanoi in Vietnam negotiates such contractual relations with 52 of the country's rural provinces in order to ensure a supply of good-quality, healthy products, thus contributing to the provinces' rural and agricultural development.

Urban waste management, for organic waste in particular, is essentially intended as a sanitary measure but can also serve to recycle organic waste into green manure, which, when used in agriculture, restores the nitrogen and phosphorus cycles broken by the use of chemical fertilizers. The distribution of poultry to inhabitants, or subsidies for henhouses provided by some municipalities, has the same objectives: reduction of urban organic waste by recycling, and promoting egg production.

Promotion of Economic Activities

Most LUGs seek to encourage or sustain economic activities that generate employment or income for the people, but also sources of tax revenue. They do so in a number of ways, creating areas for companies to set up (business zones, transport infrastructure connexions, clustering of research and innovation institutions, etc.), lending or leasing premises cheaply, adjusting business tax regimes (duty-free zones), organizing networks and services, etc. Food processing and distribution sectors, as well as catering, provide significant numbers of jobs, in particular in developing countries, where micro- and small businesses are the norm (Broutin and Bricas 2006). A number of objectives are pursued in supporting these sectors: job creation, particularly for women, often the majority in this industry; access to food for populations with little purchasing power; and even enhancement of gastronomic heritage that can help boost a tourism development policy. In European countries, some LUGs promote the development of social and solidarity economy businesses.

A characteristic form of economic stimulus involves the commercial promotion of specific products of an area, often with a view to tourism promotion, for example in the form of a special accreditation or local brand. That leverage can however also be used to foster greater connexion between the city and the countryside. For example, the municipality of Montpellier has set up a series of summer events to promote gastronomic products and specialties of the surrounding countryside to the urban population.

Social and Cultural Policy

Leverage is also frequently applied through the city's social policy, in particular as regards food aid for the most vulnerable populations.

Thus, municipal centres for social action (CCASs) distribute some of France's food aid and can assist the poorest residents with food vouchers or coupons or by paying part of the cost of children's school meals, meals on wheels for the elderly, etc.

Other food supply actions may be taken to promote the integration of certain population groups. One example is the provision of public land for people fleeing conflicts, as in Bogotá, or the creation of farmsteads by people in the process of social reintegration.

Service to the people is one lever frequently used, in particular with respect to catering, whether for children (cafeterias at schools, nurseries or recreation centres), the elderly (nursing home cafeterias, soup kitchens, meals on wheels) or vulnerable populations in general. The initial purpose of school meals has often been to encourage parents to send children to school, but they have become an assurance of food security for all, in particular in dire inequality and poverty situations. In Latin America, many cities are implementing ambitious social policies through catering, for both children and adults. The idea is not just to provide everyone, even the poorest, with a balanced meal; those who supply the school meals may, as in Brazil, be recruited exclusively from among local family farmers, thereby extending the reach of the city's social policy into the countrysides that supply it. Brazil's pioneering municipal policy on food security, in Belo Horizonte, served as a model for the Lula Government's national 'Zero Hunger' policy (Rocha 2001).

In some countries, school meals are also a means of integrating the population in social and cultural terms (Pinson 1995). Repeated sharing of the same meal at school is an integrating factor. It may however also be divisive, as is evidenced by contemporary debates on cafeteria menus with or without pork, or without any meat. There are many priorities for school meals today: improved nutrition, taste and nutrition education, learning how to fight waste, awareness of agriculture thanks to school gardens, promotion of green agricultural systems (use of organic farming products), or relocation of sources of supply.

The last-named objective is now very common in industrialized countries' urban food policies. It reflects a strong demand for proximity, both geographic and organizational, among city dwellers who are concerned with the remoteness of their relationship with agriculture and food (Chapter "Urbanization Issues Affecting Food System Sustainability") as well as the new risks associated with food system industrialization (chemical use, food artificialization, concentration of power and financialization of business, etc.).

LUGs may have public health responsibilities and apply risk prevention measures, food hygiene controls, etc. French communes, for example, have a 'hygiene' unit that comes under mayors' policing authority and may exert control over food businesses, restaurants and open-air markets, to ensure that the food provided to consumers is safe. This involves checking that food storage temperatures are complied with, that the establishments monitored are generally clean, and that staff have training in hygiene, proper attire, etc. In some cases municipalities may enter into local health contracts, under the auspices of the Ministry of Health, to be able to implement more ambitious monitoring and prevention policies.

Finally, the field of food supply may be affected by cultural policies, which may seek to take advantage of a historical heritage, or even forge an urban identity through food. They may also seek to promote mutual awareness among communities of different sociocultural origins through culinary exchanges, neighbourhood meals or 'discovery menus' in school catering.

Evaluating Leverage in Terms of Sustainability Issues

Again, using this inventory, sustainability issues can be jointly plotted with the leverage available to LUGs on a double-entry matrix, which can be used either to ascertain what types of leverage could be applied to a given food issue, or to look into what types of issue a given type of leverage might be suited to.

A matrix such as this would also raise a series of research questions about the leverage available on the different sustainability dimensions involved. As pointed out above, many of the existing levers were devised to deal with issues other than those covered by the promotion of more sustainable food systems. There are still a large number of very open-ended research questions. For example, how do we evaluate what effect it will have on the environment, on the sustainability of agricultural production patterns, on job creation, and on city dwellers' relationship with agriculture, to keep agriculture going in or around the city? Again, how can we evaluate the effects on food practices and representations of an urban food environment, and in particular commercial town planning or the offer of garden plots? How can we evaluate catering actions' effects on nutrition or intercultural dialogue? And finally, how can we evaluate the effects of food supply relocation on the environment, nutrition, or social relations?

That is not to say that answering all of these questions is a prerequisite for the development of urban food policies. Such a matrix can nevertheless at least be used for discussion purposes and to highlight the relevance and potential impacts of the available levers via collective expertise. In practice, many political actions are decided on without any real certainty as to whether they will have the desired effect, even supposing the objective to be well defined. Food policies are rarely instituted on the basis of thorough knowledge of the effects and risks of the actions to be taken. Often, they simply follow on from former practice, by inertia, with ad hoc experiments that are then tailored to the circumstances. Hence, the actual conditions under which the policies are made need attention, which is the subject of the next part.

Steps for the Construction of Urban Food Policies

Given the many city initiatives to devise food policies, i.e. a coherent set of actions to improve food supply, some institutions have felt a need to propose conceptual and methodological frameworks for such policies (Chapter "Theoretical Approaches for Effective Sustainable Urban Food Policymaking"). The goal is often to try to put all the environmental, nutritional, economic, social, cultural and food issues into one framework, *and* all food system elements over which they hope to exert control:

production, trade, catering, consumption, waste management and their various determinants. When such frameworks are developed the temptation is to urge LUGs to map their agricultural and food situation, while quantifying flows, surfaces, consumption, population; identifying and often locating all the players; listing corporate and citizen initiatives; and tracking changes for each of these elements. What makes this such a difficult exercise is the dispersal of the data, which are often kept by many different players, not necessarily working at the same scale (Brand 2015). Data may even be lacking for certain problem dimensions—one example is the environmental costs of food, a relatively new issue that has yet to be the focus of many statistical studies. It is time-consuming to do this kind of mapping and may be quite costly. It does often allow the importance of the various parts of the system to be gauged: how important urban food production is to consumption, what share of total employment agricultural and agrifood jobs account for, how serious food insecurity and nutritional diseases are, etc.; but often it has little direct usefulness in initiating actions.

The approach we advocate begins with the facts set out in the literature and observed in our practical research, namely that cities often take up the food issue for reasons other than an explicit intention to develop a sustainable food policy. There are certainly cases where that intention does exist, but it is not always, and in fact rarely is, the prime mover in current food policies. Instead, the issue arises because of other urban concerns such as health, hygiene, poverty, the environment, social movements, unemployment, etc.

In some cases, problems needing solutions will then focus attention on food, either directly (e.g. if a quality problem arises in school cafeterias), or indirectly, when food is one of the areas of activity in which solutions to a given problem may be found (to create jobs, for example, the food sector offers considerable scope for new activities).

In other cases, certain types of leverage can afford an opportunity to address urban food issues. While direct leverage is possible, through the creation of new, food-specific tools, the approach is also often indirect, as the food issue is addressed using tools originally designed for other purposes. Thus, cities have to manage waste, initially from the standpoint of health, but they then may seek to use the waste as fertilizer as an agroecological strategy in order to reduce waste storage and reprocessing needs. However, as we mentioned in the previous section, the potential for use of these means of action for food policy purposes is an area of knowledge that needs to be investigated.

Toward Policymaking on the Urban Food Issue

Starting with these three points of entry (policies, problems, levers), and the example of Bristol, where a food policy was formulated on the basis of an experience capitalization document (Carey 2011), we established a matrix of ways in which LUGs can address food: from policymaking approaches to the urban food issue.

One approach is to begin with the levers available to LUGs. Given the great variety of perceived food problems, or a desire to regain some control over food supply but without any clear identification or formulation of the problems, a first step is to look at what can be done with the levers available. In that case, the action to be taken is arrived at not on the basis of an 'objective' assessment of the situation, indicating the objectives to be achieved, but rather of the levers LUGs can use to address food issues. That approach can be described as building a 'policy of means'. Few LUG officials have food expertise, but they are experts on the means they have available, and their predisposition is to use those means in addressing the new food supply area.

In our facilitation work with municipal officials in preparation for the agricultural and food policy of metropolitan Montpellier (Chapter "Putting Food on the Regional Policy Agenda in Montpellier, France"), we often noticed that mindset, namely that the officials thought about what those levers would allow them to do to achieve different goals than they were intended for. Thus, discussions took place around ideas such as using school cafeterias to promote organic farming in the region or hosting social and solidarity economy agrifood companies within the wholesale market.

A second approach is to begin with the specific problems LUGs need to solve. Those problems are not necessarily identified through a full assessment of the city's overall food situation, following which priority actions are mapped out. Instead, the problems are generally put on the political agenda by a few influential actors. In France, the fight against food waste has emerged as a problem to be solved not in the wake of any alarming assessment, but as a moral imperative, impelled by a number of actors who were able to mobilize politicians.

In our work with metropolitan Montpellier elected officials, we discussed the question of food insecurity, a particularly thorny one for them, with a view to identifying the levers the LUG could deploy against it: cafeteria prices, access to garden plots, food aid, support for solidarity groceries, etc.

The third approach is to start with a policy proposal, conceived of as a set of actions to remedy (and/or forestall) a set of food-related problems. That involves mapping out an action framework linking the problems to be addressed with the levers available to solve them.

These three approaches to urban food policies are summarized in Fig. 2.

In practice, these approaches will be combined, interlaced, or tried in sequence. The levers are scattered through different units of the LUG and the problems to be solved are varied and may be tackled by several of them. Some units will already

Fig. 2 Toward policymaking on the urban food issue

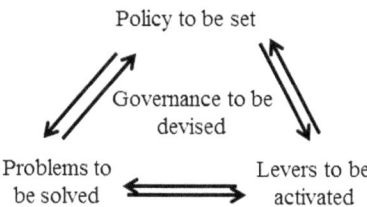

have set levers in motion when others get involved, acting on other concerns that may nevertheless have an impact on the first-comers' actions. Therefore, the construction or management of urban food policy is subject to considerable coordination and arbitration issues.

Another feature of urban food policies is that there are many initiatives and actions undertaken by civil society or the private sector that the LUG must seek a way to interact with. Hence our central focus, in this analytical framework, on the issue of food supply governance as a work in progress. That governance is indeed the central process, wherein policy choices, problems to be solved and levers to be pushed must be reconciled.

An Iterative and Integrative Methodological Approach

What consequences will such a conception of food supply policymaking have at the urban scale on support operations and research? In considering how to capitalize the support to be provided to metropolitan Montpellier's agroecology and food policy, the following three methodological principles may be proposed:

Principle No. 1: combine the three elements. The support to be provided is not a 'turnkey' operation. It will embody a procedure that combines the three elements to be included in food supply governance. While not in contradiction with existing approaches based on expert assessments, it will seek to integrate these into a comprehensive approach that will be fleshed out by the actors undertaking it, not laid down *a priori* by researchers or design offices.

Principle No. 2: provide tools for exploration of each element. Many kinds of knowledge will need to be generated to support urban food policies; using our design framework we can discern various types of assessments to perform.

- Problems: assessments to ascertain and understand them. Here, classic expert assessments will be needed in areas of food policy where the action that should be taken cannot proceed for lack of knowledge. Perhaps we have identified or are aware of a problem but its causes are poorly understood. In such a case an assessment establishing the causes or an investigation may make it clear how to proceed.
- Levers: assessments to control them. Such assessments are often lacking. While we may know how a given lever affects the problem it was designed for, it is much less clear how it will affect other problems, or what its systemic effects on food supply will be. What is needed in that case are assessments and joint innovation approaches that help actors to formalize work procedures that both take account of the objectives and embody suitable ways and means of acting.

- Governance: assessments for positioning and evaluation. One requirement here is for models of the food system that can give actors an objective idea of the context in which their actions will be taken. For this purpose, the available outlines showing the various components of the food system and the issues involved, such as the one produced for metropolitan Montpellier, may be useful (Fig. 3). Another requirement, however, is to equip these actors with a way of evaluating the governance actually being undertaken by means of indicators of how far the objectives have been met and with what effects. Evaluation, so conceived, is the outcome of a learning process in which scientific knowledge and contextual knowledge are combined (Rey-Valette et al. 2014).

Principle No. 3: support governance. The international literature and our local experience (Chapter "Putting Food on the Regional Policy Agenda in Montpellier, France") teach us that urban food governance develops over the long term. Its objectives are ever-changing, the actors involved in achieving them change during the process, and the achievements and their knock-on effects bring about lasting readjustments. This kind of governance is adaptive and specific knowledge is required to direct it. Human and social science research can play a supporting role in the monitoring of governance, employing tools that 'map' the actors involved and, of these, those who are or are not 'enrolled' in the sociotechnical process (according to Michel Callon's sociology of translation 1986).

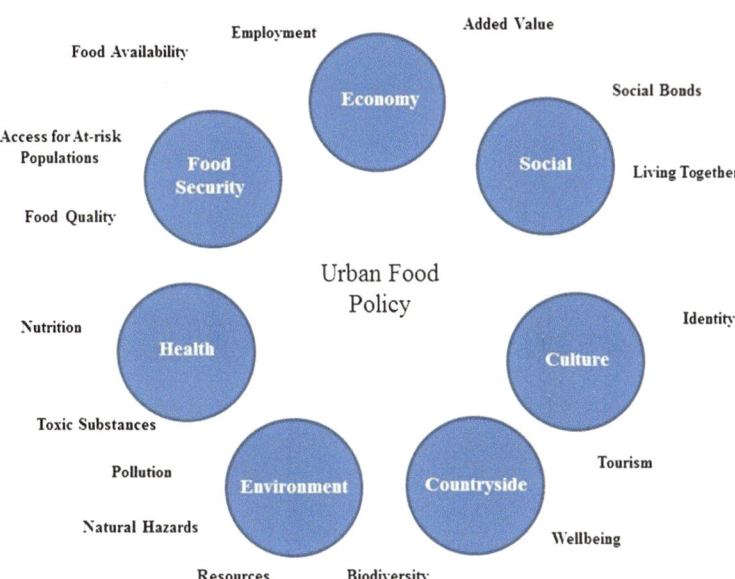

Fig. 3 An outline of the possible purviews of urban food policy

A further requirement is for internal analyses of the strategies that players develop within organizations involved in governance (according to the 'sociology of organizations' of Philippe Bernoux et al. 2015).

The involvement of researchers in such processes implies that these flows of knowledge must not be seen only as reports of observations at a remove, but that interactions must be designed and directed throughout the public policymaking process. The design of procedures to support action research (Liu 1997) can be used to frame cooperative relationships between researchers and actual practitioners. Using such designs, the three research figures/postures may be combined to ensure a lasting partnership between researchers and actors on the ground (Soulard et al. 2007; Petit and Soulard 2015): expertise on targeted assessments, facilitation aimed at integrating disseminated knowledge during decision making, and the action of an evaluator seeking to endow the policy with an ongoing evaluation of the governance process.

Conclusion

What kind of conceptual framework can we develop that will cover at once the environmental problems posed by cities' food supply systems, nutritional problems or those occasioned by living together in the city? Rather than add one more new proposed scheme to the many that already exist, each of which seeks to map the complexity of urban food systems (Chapter "Theoretical Approaches for Effective Sustainable Urban Food Policymaking"), we resolved to explore an element that is little studied in the literature: the levers available to local urban governments to influence food supply. By identifying these levers and matching them with the food challenges caused by urbanization, we have been able to map certain research issues. In looking at each of the levers, we can investigate its real or potential effects on each of the potential problems to be solved. At that point, a specific conceptual framework needs to be developed for each of those issues. A study of the effects of commercial town planning on the sustainability of food practices cannot be done in the same context as one that looks into the effects of agricultural recycling of urban waste. Yet another framework will be needed to study the job creation impact of a food production relocation policy.

Accordingly, rather than propose an integrative scheme based on an expert's or planner's vision, we have sought to recognize the many different paths of policy development required by food issues. As we learned from our experience of supporting food policies, that approach can cater to actors from many different areas of activity, with often different and even divergent interests. If we as researchers are to support these actors in the construction of their policies, we need to acquire the ability to test the hypotheses that emerge at the interface of the new problems to be resolved and the levers available to them. Further, we need to cooperate with them to come up with new approaches suited to new challenges.

References

Bernoux P, Amblard H, Livian YF, Herreros G (2015) Les Nouvelles Approches sociologiques des organisations. Seuil

Brand C (2015) *Alimentation et métropolisation: repenser le territoire à l'aune d'une problématique vitale oubliée.* Doctoral thesis in Geography, Université de Grenoble, UMR Pacte, 656 p

Broutin C, Bricas N (2006) Agroalimentaire et lutte contre la pauvreté en Afrique subsaharienne. Éditions du Gret, Paris

Brunori G (2015) Alternative food networks as drivers of a food transition. In: Viljeon A, Bohn K (eds) Second nature urban agriculture. Designing productive cities. Routledge, Abingdon

Callon M (1986) Some elements of a sociology of translation: domestication of the scallops and the fishermen of St Brieuc Bay. In: Law J (ed) Power, action and belief: a new sociology of knowledge? Routledge, London

Carey J (2011) Who feeds Bristol? Towards a resilient food plan. Bristol Green Capital and Bristol City Council, Bristol

Cohen N, Ilieva RT (2015) Transitioning the food system: a strategic practice management approach for cities. Environ Innov Soc Trans 17:199–217

Epstein L, Raja S et al (2012) The built environment moderates effects of family-based childhood obesity treatment over 2 years. Ann Behav Med 44(2):248–258

Hellec F, Barataud F, Martin L (2013) Protection de l'eau et agriculture: une négociation au long cours. Natures Sciences Sociétés 2(21):190–199

Jarrige F (2013) The comeback of the food issue in the city of Lausanne: a challenge for the coherence of local policies. In: 5th AESOP conference on sustainable food planning. 2013-10-282013-10-29, Montpellier, FRA

Lascoumes P (2010) L'Éco-pouvoir: environnements et politiques. La découverte, Paris

Le Strat A (2008) Pour une eau du robinet de qualité. Un volet du développement soutenable dans une grande ville. Annales des Mines – Responsabilité et environnement 1(49):36–41

Liu M (1997) Fondement de la recherche-action. L'Harmattan, Paris

Petit S, Soulard CT (2015) Rencontres par-delà les frontières: l'analyse du partenariat chercheurs-acteurs dans le programme de recherche PSDR. In: Les chercheurs ignorants (eds) Les recherches-actions collaboratives. *Une révolution de la connaissance.* EHESP, Rennes, pp 93–100

Pinson D (1995) Générations immigrées et modes d'habiter. Entre repli communautaire et fusion transethnique *Les Annales de la Recherche Urbaine* 68–69:189–198

Raja S, Li Yin L et al (2010) Food environment, built environment, and women's BMI: evidence from Erie County, New York. J Plan Educ Res 29:444–460

Rey-Valette H, Chia E et al (2014) Comment analyser la gouvernance territoriale? Mise à l'épreuve d'une grille de lecture. Géographie, économie, société 16(1):65–89

Rocha C (2001) Urban food security policy: the case of Belo Horizonte, Brazil. J Stud Food Soc 5(1):36–47

Shove E, Pantzar M, Watson M (2012) The dynamics of social practice: everyday life and how it changes. Sage, London

Soulard CT, Compagnone C, Lémery B (2007) La Recherche en partenariat: entre fiction et friction. Nat Sci Soc 15(1):13–22

Viljoen A, Wiskerke JS (2012) Sustainable food planning: evolving theory and practice. Wageningen Academic Pubication, Wageningen

Warde A (2005) Consumption and theories of practice. J Consum Cult 5(2):131–153

Yin L, Raja S et al (2013) Neighborhood for playing: using GPS, GIS, and accelerometry to delineate areas within which youth are physically active. Urban Stud 50(14):1–18

Putting Food on the Regional Policy Agenda in Montpellier, France

Laura Michel and Christophe-Toussaint Soulard
with Montpellier Méditerranée Métropole (Box Contributor)

While the food supply question virtually disappeared from the agenda of the cities of the developed world after World War II, a growing number of cities in the developing and developed world alike are now giving it their attention (Chapter "Cities' Strategies for Sustainable Food and the Levers They Mobilize"). That renewed focus has been brought about by a food supply discourse that seeks to promote a sustainable urban policy (Pothukuchi and Kaufman 2000). In that context, food supply has come to be at the heart of a new category of public urban action—sustainable development—that has now been widely undertaken in local public action (Béal et al. 2011).

It appears, however, that the types of policy pursued are highly variable depending on which cities are looked at. Some propose policy instruments focused on the food-health link, as in Toronto, while others are concerned with the food cycle and target waste management, as in San Francisco (a 'zero waste' city). City administration of food issues is very recent in France (Perrin and Soulard 2014). Though a few pioneering cities did adopt food policies, often agriculture-based at first, this level of government only recently received legislative authority from the State, in the form of the territorial food projects established under the 2014 Orientation Act.[1]

In this chapter, by tracing the experience of metropolitan Montpellier, which in 2015 devised an agricultural and food policy, we explore the pathways being taken as the urban food issue emerges. How is it put on the city's political agenda? How does it take shape as a public issue? In other words, how does food as a social fact become an object of concern and debate and possibly public action? How does the

---

[1] Act No. 2014-1170 of 13 October 2014 (*Loi d'avenir pour l'agriculture, l'alimentation et la forêt*).

L. Michel
UMR Cepel, Montpellier University, CNRS, Montpellier, France

C.-T. Soulard (✉)
UMR Innovation, Montpellier University, INRA, CIRAD, Montpellier Supagro, Montpellier, France
e-mail: soulard@supagro.inra.fr

© The Author(s) 2019   123
C. Brand et al. (eds.), *Designing Urban Food Policies*, Urban Agriculture,
https://doi.org/10.1007/978-3-030-13958-2_6

food issue end up taking institutional form as a food policy? At the local level, what knock-on effects does that have?

To answer these questions, we shall be making three general assumptions that are likely to have an impact on the urban food agenda and the way the food issue is framed. Our first assumption is: while sustainable food policies may seem new, they take their place within existing institutional configurations, affecting both the potential emergence of a new agenda issue and the way that issue is framed. The food issue does not crop up spontaneously on the political agenda. It must be put forward by actors—entrepreneurs—who socially construct it as a public issue for the city. Indeed, the literature on public problems underscores the role of policy entrepreneurs (Becker 1963; Cobb and Elder 1972) in the emergence of new problems in the public area, thanks to cognitive work to define the problem and a mobilization effort. There is a wide variety of issue entrepreneurs in the area of food: thus, actors such as States, the European Union (EU), FAO, the scientific community, social movements and networks of cities, have proposed food action models to be pursued within and between urban areas. While the models do propose innovative supply arrangements, which can spark public debate on the food issue, and provide some framing of ways to deal with it, they need to be taken up and interpreted at the local level by actors who can facilitate a connexion between the processes of mobilization around problems and their management under public policies. These actors are then called policy entrepreneurs (Kingdon 1984; Guiliani 1999). In some cases, the local food movement plays a central role in this interpretative process. Our perception in Montpellier, in contrast, is that the decisive role has been played by local State bodies, certain local authorities and the scientific community. Finally, the transformation of existing institutional configurations can offer a window of opportunity for a new problem like food supply to be placed on the political agenda (Kingdon 1984). Territorial restructuring of the State's role leads to responsibility being delegated to (but also caught by) city governments for these new issues. Similarly, times of institutional change or political coalition renewal are also conducive to the adoption of innovative policies—or policies perceived to be so. Our third hypothesis, then, is that the emergence of a new agricultural and food policy is bound up with changing institutional and territorial configurations, and especially the reshaping of the sociopolitical relationships between the city and its interstitial farm or garden areas as well as between the city centre and its rural outskirts.

By retracing the process of food policy development in Montpellier we can gain an understanding of how the food issue was put on cities' agenda in the French context. In 2015 an agroecology and food policy was voted in by Montpellier Méditerranée Métropole and came into effect in its 31 constituent municipalities. The food issue was added to the agenda in a legislative, electoral and regional context that needs some explanation. The first part shows under what conditions a city was able to address the food issue. Then, in the second part, we explain how that was done, by whom, and by means of what framing effects with respect to the policy proposal passed in 2015. In the third part we look at the effects the policy has had on local institutional configurations, relations between the constituent municipalities of Montpellier Méditerranée Métropole, and the emerging governance modes there.

Putting Food on the Regional Policy Agenda: An Agricultural Prism

Gilles Pinson defines the urban policy agenda as 'the set of social facts that count as public problems, controversial and debatable at city level and calling for public action at the same level' (Pinson 2006: 620). In the case of Montpellier, the food issue did not arise directly as such. It was the agricultural issue that was first raised by local actors as a public action item, in the years from 1990 to 2000 when wine-growing was in crisis. The action first developed was support for local wine production and, subsequently, protection of farmland from development in the metropolitan Montpellier area. Paradoxically, the prime movers in raising the food issue in local public discussions were the State—in particular through the national health nutrition plan (PNNS), then the national food programm (PNA)—but also the Agropolis scientific community. As it turned out, the 2014 municipal elections gave some new political actors an opportunity to put the food issue on the metropolitan agenda.

Emergence of Agricultural and Food Insecurity Problems

Whereas under the Fordist model the State, through planning, dominated the organization of local production systems, the State's disengagement as of the mid-1980s—a boon or a constraint?—opened up new areas in which cities could act (Pinson 2009). Certain problems previously dealt with by the State—such as the agricultural issue—then cropped up on local agendas. In the urban area around Montpellier, the first initiatives of the district, then of the Montpellier conurbation community, were to promote local vineyards. These initiatives were supported by elected officials of small periurban municipalities, who advocated for a wine-growing region dominated at that time by monoculture. Thereupon, city dwellers' demand for quality local products led to the development of shorter food supply chains, taking the form of farmers' markets or points of sale on the farm or in town. While wine-growing had to that point been supported by the State, then the European Union, part of the industry then pivoted toward the production of quality wines, and other winegrowers that had been producing mass-market wines permanently grubbed up their vines. As a result of these changes in the periurban vineyard landscape, production diversified in response to the needs of farmers near Montpellier who switched to market gardening or grain crops and to a certain demand from urban consumers (Perrin et al. 2013; Scheromm and Soulard 2018).

Local actors' other main focus, as regards the food issue, was a social one. The economic crisis spurred social inequalities in the cities of the developed world, and in particular food insecurity. The shock was much sharper in English-speaking countries, where cities in crisis in the 1980s were basically left to their fate (Stoker 1991). In this context, the issue of access to food emerged as one of the baselines for urban food planning policies (Chapter "Theoretical Approaches for Effective

Sustainable Urban Food Policymaking"), as in Toronto for example (Friedmann 2007). Conversely, in France, the European Union and the State, through the European Food Aid Programme (PEAD) and the national food aid plan, together with food aid associations working locally, played a central role. However, the scope for State action appeared limited, on the one hand by growing local social problems, and on the other by the transfer of national jurisdiction to the departments. And indeed the Department of Hérault and the community centres for social action (CCASs) leveraged their social jurisdiction to expand their food aid efforts. In Montpellier, innovative actions, such as short supply chain support for the *Restos du Cœur*, were undertaken in conjunction with the wholesale trading centre (MIN) and INRA and with the support of the Regional State Division of Food, Agriculture and Forestry (DRAAF) (Le Velly and Paturel 2013). It should be noted that the department and the municipalities are in charge of school catering. In that context, their pricing policies, which take family size and income into account, contribute to better access to food for children. However, catering has not really been thought through by local actors as an instrument for a comprehensive understanding of the food issue. Only the municipality of Grabels stands out, with its more comprehensive approach to food supply, which combines the creation of a short supply chain market—implemented with INRA support[2]—with the use of catering as leverage to promote local product supply, and action to provide farmers with land.

Hence, the food issue has not come out of left field. As public action has become more territorial, local governments have had to take on problems which, while not at first defined as 'food' problems, are concerned with such related issues as agriculture and social insecurity. However, even though some municipalities in the metropolitan area, like Grabels, have developed a more comprehensive approach to the food issue, before 2010 it remained very fragmented and siloed among local actors. Moreover, the metropolitan administration as such and most of the municipalities remained relatively passive.

Role of Regional State Authorities and the Scientific Community in Handling the Public Food Supply Issue in Montpellier

A variety of actors are involved in the emergence of the public food supply issue in cities. In defining this new issue in the public space, a contrast may be drawn between bottom-up dynamics, originating with civil society, and top-down ones, for which administrative or political elites are responsible (Stierand 2012), or indeed between convergent and competitive initiatives.

Elites sometimes take advantage of the social movement to develop and implement their policy. Actors from academia very often play an important role in spreading

---

[2] Creation of a collective label (IciCLocal) jointly filed with the INPI by INRA and the city of Grabels.

these ideas. Such is signally the case of the Anglosphere's food planning movement, in which researchers are heavily involved (Derkzen and Morgan 2012; Brand 2015). It is true, too, of Montpellier, where the scientific community of the Agropolis research centres, which works on problems of agricultural and food sustainability, has played an awareness-raising role with politicians through a number of research projects undertaken in partnership with local authorities over the past decade.

And lastly, while international trade liberalization negotiations have cast doubt on the role of the State as a protectionist force and provider of agricultural subsidies, that role has been reconfigured rather than abolished. In particular, the food supply issue has been taken up by the State,[3] which has developed an incentive policy in the form of the national food plan (PNA), which seeks to encourage territorial actors to involve themselves in the issue of food, its quality (nutritional and heritage-linked) and accessibility (Bonnefoy and Brand 2014). While DRAAF mobilization is variable depending on the region of France concerned, the Languedoc-Roussillon DRAAF appears particularly active in implementing the regional food programme (PRA). In spite of the PNA's constrained financial resources, the two project officers of the DRAAF nutrition and food supply quality unit have helped to stimulate local debate on the food issue, focusing on areas of activity that frame the issue primarily in terms of agriculture and food: these include short supply chains, the 'fruit for recess' project and other educational activities, as well as heritage issues and allotment gardens; and, on the other hand, the issues of catering and food aid. The two project officers have been responsible for devising and disseminating best practices in food policy across the region. To that end, they have sought out, gathered, sorted and organized (on index cards) a whole range of information on food initiatives under way in the region, relaying information from one to the other and helping them to network: local/organic food supply for cafeterias, territorial food project, land programme, etc. DRAAF too is active in popularizing these initiatives, organizing events where information gathered on current experiences is presented and shared, then disseminated to partners through a variety of channels. Lastly, DRAAF is involved in training, and seeks to share these best practices in that area too. In short, even though it has no formal hierarchical role empowering it to exert direct control over local authorities, it seems that in practice, after 5 years of field work, DRAAF is actively contributing to the emergence of food as a public issue (Michel et al. 2014).

Thus, in the early 2010s, while the food issue continued to be addressed in a piecemeal fashion through various poorly coordinated sectoral approaches (agriculture, school catering, food aid), there was an emerging focus on food supply as a discrete urban issue. At first it was a focus mainly of the State and the Agropolis scientific community, but as the conurbation underwent institutional and political changes, the issue made its way onto the local political agenda.

---

[3] In 2009, through the Ministry of Food, Agriculture and Fisheries (MAAP).

Changes in the Regional Political Configuration: A Window of Opportunity for the Food Issue

The appearance of multi-level governance across Europe (Marks 1996) also led to changes in the way agricultural and environmental issues were dealt with at the institutional level. In Europe, reforms to the Common Agricultural Policy made for greater involvement of regional governments in handling agricultural problems and a new environmental focus (rural development, agroenvironmental measures, management of the wine crisis). At the same time, the various territorial and administrative reforms redistributed local jurisdictions. In France, the rise of intercommunality resulted in profound changes. The 2001 transformation of the—very urban— Montpellier district into a much broader conurbation community changed the socio-economic balance within the urban area. The mobilization of newly integrated rural municipalities and DRAAF support succeeded in putting the agricultural issue on the new intercommunal agenda thanks to the preparation of an agricultural assessment upstream of the Metropolitan masterplan, named 'SCoT' (Jarrige et al. 2006). Finally, as a result of the Rio Conference of 1992, local Agendas 21 have multiplied. Everywhere their implementation has fuelled a discussion on 'green' areas in the city and its outskirts, with new virtues being found in agricultural areas. In Montpellier, the challenges of managing natural and agricultural areas, on the one hand, and periurban agriculture on the other, were placed on Agenda 21 adopted in 2011, assigning two project officers the job of investigating the 'agricultural problem' part-time. Thus, the environmental argument helped legitimize the agricultural issue within metropolitan Montpellier. Some natural or agricultural areas, up to that time considered only in the light of land reserves for urbanization, suddenly acquired an environmental value that made them central to sustainable urban development. Since the mid-2000s, collective gardens and agricultural parks in particular have become highlights of the urban projects of Montpellier Méditerranée Métropole, even where the food issue as such is not specifically mentioned.

The 2014 election of a new team to administer the conurbation seems like a window of opportunity (Kingdon 1984). The fact that the new Mayor of Montpellier, when also elected President of the conurbation, emphasized his desire for political innovations in response both to the changes in local society and the aspirations of the executives in charge of its policies, opened the door to addressing new issues but also embarking on new forms of governance. Indeed, the new political balance within the metropolis has given a boost to an official who is committed to agricultural and food issues. The President of the conurbation and future metropolis has to deal with its rural municipalities and also the political balancing act that placed the Greens in a favourable position. Thus, an innovative project combining agriculture and ecology has been entrusted to an official - a woman- from civil society who embodies renewal of the political establishment.

Construction of a Regional Agroecology and Food Policy

Urban food policy development is not part of any defined regulatory schedule. In France, agriculture, like food, is not something for which local, communal or inter-communal governments are responsible.[4] Similarly, national and European policies governing these sectors offer simple incentives, not binding on the local authorities and relatively recent developments. In particular, the State seeks to encourage actions it can no longer take directly, by enriching local initiatives. Accordingly, we find ourselves in a situation where, if local urban governments decide to embark on such projects, they do so mainly on their own initiative. As a result, relevant frameworks need to be found, or devised, for the actions to be taken. Their particular context is innovation, which gives particular importance to the knowledge that needs to be leveraged for action at the local level, and to collaborations that can be established with researchers. The essential knowledge for the actions to be taken pertains both to the particular field, the urban food system, and to the skills, tools and means available to local authorities to act locally on the food system and guide its development. Montpellier's experience shows how that endeavour has unfolded and the role to be played by knowledge mobilization, with some help from researchers, in formulating an 'issue' with which local elected officials can identify.

The Spark: Political Renewal and Administrative Reform

Food policy development in Montpellier has come about as the result of a combination of several changes in local political life. First, a new political team took charge of the conurbation community, seeking a clean break with the past. As public administration underwent reorganization, the challenge for Montpellier was to be recognized as the 11th Metropolis in France, joining a list of ten metropolitan areas or urban communities already chosen by the State for designation as 'metropolises' as of 1 January 2015. That challenge obliged the President-elect to gain the consent of the 30 municipalities making up the conurbation as well as that of the City of Montpellier, whose new mayor he was. He was also called upon to persuade the largest possible number of intercommunal authorities in the region to sign a cooperation pact with Montpellier focusing on a few strategic development areas. Pivotal in this endeavour to enlist the surrounding areas' support were the issues of agriculture, food and rural life. Those issues are not being coordinated only by the President himself, who relies instead on the Vice-President he has appointed to take charge of them. She is the newly elected Mayor of one of the smaller municipalities in the conurbation, an agricultural engineering by training whose background is civil

---

[4]A state of affairs now subject to change: the Bonnet report (2016) proposes that communities be invested with responsibility for food.

society (environmental sphere). Her agroecology and food responsibilities[5] mean that she will take the opportunity to launch a public policy as innovative and comprehensive as she can make it.

Meanwhile, the matter taking up most of the conurbation's time and effort is the launch of the assessment of the second version of its SCoT, in which the terms of reference of the tenders for the hiring of design offices are defined. The Vice-President therefore decided to take advantage of that requirement to obtain assistance from researchers in drawing up the specifications. Having worked in Montpellier's agricultural training college, she got in touch with a research team at INRA whose work she was familiar with—a study of periurban agriculture undertaken for the agricultural assessment under the first SCoT (Thinon et al. 2003).

Knowledge of the Urban Food System: Partial, Haphazard

As a corollary of their land use and urban planning expertise as well as economic and social development, local authorities rarely have technical resources in the agricultural and food sectors. The Montpellier conurbation community, which undertook to devise a SCoT masterplan with emphasis on agricultural and natural structuring in the development of its territory, had not initially acquired any expertise in that area. Since Agenda 21, two task officers have been given the following part-time assignments: for one, inclusion of agriculture in communications on economic development; for the other, coordination of the agri-parks component included in the SCoT zoning. Some assessment work was financed also through an agreement with the Chamber of Agriculture.

When, wishing to launch her policy, she contacted the INRA research team, the Vice-President's immediate focus was the renewal of the SCoT agricultural assessment. However, her first contacts with a researcher led her to widen the scope of her intended policy, realizing, thanks to the research team's presentation of examples of policies implemented elsewhere, in English-speaking countries in particular, that the food issue covers a far broader range than just agriculture. That wider policy scope was by no means alarming to the community's officials, as they realized that the food issue was not something foreign to their own concerns but rather shed a different light on areas in which they were already active. As the meetings progressed, nine divisions of the intercommunal government apparatus, out of 15, were deemed to be directly impacted by the issue. How then should a policy affecting services right across the board be correctly designed? How should the area's other vice-presidents and other officials be asked to pitch in, with only the one vice-presidency specifically designated as having responsibility for food?

These initial thoughts led to the idea that the policy to be developed needed to come from the stakeholders of the whole conurbation and its 31 municipalities, be

---

[5] The contours of her vice-presidency, which initially consisted of 'SMEs, crafts, agriculture, rural life and traditions' will in 2015 be refocused on 'agroecology and food'.

they elected officials or services officers. Technically, the commissioning of specifications under the SCoT masterplan, in the form of works management assistance, was the operational basis of the emerging contract with the research team, but everyone's common objective from the start was to formulate goals on the basis of which a comprehensive public policy could be put in place. The work schedule was tight, however—6 months at most—as the policy was to underpin the future metropolis project. Because of that constraint, a pragmatic approach was taken, based on discussion of possible ways forward rather than a formal expert opinion. The discussions proceeded by pooling knowledge on what a 'food policy' would mean in Montpellier by means of workshops with the mayors of all 31 municipalities as well as other elected officials and services officers,[6] drawing inspiration from the approach taken by the experimentation and research group on development and local action (GERDAL) (Darré 2006). The researchers organized the workshops and participated in them, to provide documentation and facilitate the debate, as facilitators rather than experts.

The participants were presented with a brief 'agricultural and food portrait of the territory' to stimulate dialogue on the characteristics of the local agricultural and food system. The first workshops pointed up the primacy of farmland in the debates among local officials, for whom the question of what to do with periurban land was a primary concern. The food issue was less present in the workshops, where the point of view of the periurban elected officials was stressed, apart from the subject of school catering, which was also within the municipalities' purview. As the discussions went on, the researchers' contributions winkled out participants' main areas of concern, such as farmland, agricultural facilities, canteens or farmers' markets. Other areas were dealt with more quickly (environment and tourism) or even ignored as being perceived to be outside the communities' remit (health, trade) or too little known (link between insecurity, health and food supply).

Realizing how little they knew of the urban food system made participants aware of the areas in which they were active and those where they had no role. They were helped to explore areas more remote from their everyday by a presentation the researchers gave on a more global view of the food system, one that also featured innovative actions, little-known initiatives, but also discrepancies between the various municipalities, in particular compared to those engaged in pioneering efforts in some areas of food policy. The idea thereby emerged, too, that while food supply was a new subject of public policy, numerous initiatives did already exist; and while these were quite various, they did constitute a possible basis for a 'unifying' policy that would to some extent constitute a revelation or synergization of actions already under way.

The workshops not only gave participants a better grasp of the food system, it also impressed upon them that the policy to be devised would require both new actions and an effort to coordinate existing ones. The action knowledge imparted

---

[6]An analysis of the organization chart of the conurbation community, now a metropolis, established that nine of the 15 branches were directly concerned by the agricultural and food policy (P2A). The unit heads and elected officials concerned were then invited to the workshops with the mayors of the 31 municipalities making up the metropolitan area.

was therefore of several kinds: knowledge of the purpose, of who should take action, and of what they should do.

From Assessment to Policymaking: Framework Policy and Adaptive Governance

The Montpellier conurbation community received metropolis status on 1 January 2015. Montpellier Méditerranée Métropole came into being, and as a result some powers previously exercised by the municipalities (e.g. urban planning) were redeployed to the intercommunal level. At the same time, the community's vitality won it the coveted French Tech certification given to cities that have fostered the creation of innovative start-ups. The situation was highly conducive to the preparation of the new agroecology and food policy. Even before the research team handed in its report, entitled *Preparatory Study of an Agricultural and Food Policy* (Soulard et al. 2015), the community was beginning to communicate, through press conferences and interviews in local newspapers. The city seized on the food issue to amplify its territorial marketing message. The issue gained all the more prominence owing to the political situation, a few months before the regional elections, as well as the merger of the Languedoc-Roussillon and Midi-Pyrénées regions. In a bid to compete with the city of Toulouse, the President of Montpellier Méditerranée Métropole made more and more overtures to the region's intercommunal authorities, creating a 'territorial parliament' bringing together some 50 of them.

It was at this point that the new VP in charge of agroecology and food met once again with the metropolitan area's elected officials to hammer out the overarching themes of the future metropolitan policy: six workstreams and three cross-cutting themes, all focused on five goals. The discussions dealt with the actions to be taken, the means to be deployed and the implementation timetable. The VP addressed participants' concerns by emphasizing the need for high-profile actions to be taken quickly even while maintaining a long-term focus, and so proceeding in stages, which would be dictated by the political agenda. The project was presented in community council and adopted on 25 June 2015. A 'P2A' team (on agroecology and food policy) took shape at Montpellier Méditerranée Métropole, made up of four officers belonging to the Economic Development and Operational Land Management services, as well as a City of Montpellier officer in charge of school catering whose mission was to bring his ideas to the attention of the other municipalities of Montpellier Méditerranée Métropole. A few actions were set in motion straight away. At the same time, action programmes were under development on the various themes, with separate committees being formed to have the stakeholders draw up task sheets. Completion of the preparatory study also marked the end of the working relationship with the researchers, which had been quite intense during the workshops.

A second phase then got under way to help follow up and evaluate the policy. A multi-year agreement on cooperation with the researchers has been signed.

While the goals of the policy now being undertaken do cover the main elements of the food system, in practice its priority themes are focused on a few actions based on resources actually available to Montpellier Méditerranée Métropole, e.g. its public lands or its municipalities' canteens (see Box 1). Hence, the emphasis is on the agricultural restoration of available land by settling farmers on it, with a variety of production systems suited to urban demand: small market gardens and more specialized farms, able to supply mass catering with products sold through short supply chains. The governance to be pursued is meant to be pragmatic and adaptive. While political considerations would dictate some urgency in taking action, the limited means available to the team being formed will require them to moderate their ambitions. Certain themes are not really addressed in the first set of actions—to wit, health measures and food availability for poor households—but that issue is left for another day. A frame-

---

Box 1: Metropolitan Montpellier's Agroecology and Food Policy
Source: Montpellier Méditerranée Métropole

The P2A is structured around five goals, each of which requires an initial assessment so that baseline indicators can be established for its evaluation. Those goals also serve to identify and indeed select the actions to be supported by Montpellier Méditerranée Métropole. They are: To provide healthy locally produced food to the masses. To support the agricultural and agrifood economy and employment therein. To preserve the landscape heritage and natural resources. To limit GHG emissions and adapt to climate change. To promote social cohesion by caring for the link with nature and the relationship between city and countryside. Six workstreams form the operational heart of the P2A. For each, a goal is defined, broken down in a set of task sheets. The key food system stakeholders are not the same for each stream. The resulting actions are defined as development proceeds.

Workstream	Targets	Actions (excerpt)
To consolidate the fabric of small farms involved in direct sales	Small-scale farmers, committed consumers	A guide to sales outlets for local products A 'resource farm' as a place of excitement and experimentation
To promote local supply in the city, in particular in catering	Specialized farms, consumers	Expand public procurement support channel structuring consolidate and modernize the MIN
To promote the diversity of the area's iconic products and expand agri- and oenotourism	Oil-producing orchards, vineyards	Support for product merchandising Development of an agritourism strategy

(continued)

Box 1: (continued)

Workstream	Targets	Actions (excerpt)
To support innovative enterprises in agrifood and agriculture provisioning fields	Upstream to downstream companies	QualiMed competitive cluster green tech and agro-tech nurseries
To mobilize citizens around food supply and the producer-consumer link	Consumers, 'gardeners'	Écolothèque recreation Centre gathering of initiatives call for ideas (food practices, fight against waste, urban green space, etc.)
To forge a consistent approach to the integration of agriculture into development projects	Developers, communities	Coordination with SCoT masterplan and PLUi zoning code for urban agriculture

The governance instituted is meant to be pragmatic. The point is to make do with what exists and to act on the food system to effect change. The approach is based on building trust between all those involved.

work policy has actually been formulated, but not yet accompanied by any follow-up and evaluation scheme. Governance is being put in place as we go along, but is subject to a number of constraints due to political imperatives and Technical Services' institutional organization.

Organizational, Political and Territorial Reconfigurations

Politically, the fact that P2A was adopted by the metropolitan Council gives it great legitimacy; and the workshops from which it originated attracted many elected officials and administrative staff, who then got involved in developing the project. Finally, the project has the support of the President of Montpellier Méditerranée Métropole. Its implementation does however pose a few challenges, in particular because of the necessarily cross-cutting, multi-actor and multi-level nature of its governance.

One immediate question mark is its political backing. During the workshops, it was mainly those elected officials with some awareness of food and agriculture issues—mayors or deputy mayors of the municipalities of Montpellier Méditerranée Métropole—who got involved. Now that the main principles have been endorsed, implementation will depend on a solid political coalition able to validate a number of choices that will have more direct consequences, for example in terms of land or catering. Rather than attempt to forge a majority on the metropolitan council, the first task will be to win the municipalities' mayors over to a collective position suited to the way Montpellier Méditerranée Métropole is governed—that is, a form of metropolitan governance drawing its inspiration from 'neo-regionalism' (Lefèvre

1998), not based on a top-down approach but rather on the quest for consensus between multiple government actors. In actual fact, in the operations of Montpellier Méditerranée Métropole, the mayors of each municipality have a prominent place in negotiating intercommunal policies, which therefore are consensus-based; in Montpellier that consensus is ratified by the signature of the 'Confidence Pact'. Mobilization around the P2A must therefore be done primarily at the intercommunal level, as it provides for each municipality to list the workstreams it finds useful and approve a communal P2A to be put to a vote by the municipal council.

At the territorial level, that requires the Vice-President in charge of the policy to forge a consensus among municipalities with very different profiles. Indeed, agricultural and food issues are very differently perceived from one municipality to another, some large urban ones being focused on city dwellers' concerns, e.g. nature and food, while small periurban municipalities have instead an agricultural, rural profile. Existing initiatives show that these municipal profiles afford many different ways of looking at the food issue. For example: urban collective gardens or food aid in Montpellier; short supply chains, health and rural life in Grabels; agricultural and environmental land policy in Lavérune; or the management of the agricultural built environment in Saint-Geniès-des-Mourgues. A central concern of the P2A is in fact openness to further town-and-country cooperation regarding food supply channels for catering. One might speculate that this could lead to a real 'interterritoriality' as defined by Vanier (2003), i.e. the setting in motion of cooperation on food supply channels via agreements between private actors, pooling of public facilities, shared governance structures, etc. If municipal versions of the P2A and cooperation with communities beyond Montpellier Méditerranée Métropole can be achieved, that would be a good sign for the makeover of town-and-country relations that will be pivotal in building urban food governance.

Of course, the food issue involves responsibilities that are not necessarily those of Montpellier Méditerranée Métropole. For example, catering is essentially under regional and departmental jurisdictions (for high and secondary schools) or the municipality's (in the case of kindergartens and primary schools); and the municipalities, having been stripped of many strategic responsibilities by enhanced metropolitan intercommunality, are naturally very keen to maintain their prerogatives in that area. Agriculture, on the other hand, is a jurisdiction shared between many different territorial levels: European, national, regional and departmental. And food insecurity, finally, is under the jurisdiction of the department and the municipalities' CCASs (social education centers), given their social affairs mandate. What Montpellier Méditerranée Métropole can legitimately be involved in, therefore, remains to be determined, as governance will have to link a number of areas of multi-level cooperation. These findings are in line with the work of the new regionalism on metropolization (Lefèvre 1998; Brand 2015).

Lastly, at the organizational level, the P2A is currently an innovative policy championed by a team of four persons. Its cross-cutting nature means that it will require a number of long-established administrative sectors to work together. The classic challenge, here, is the siloing that has long been observed in work on organizational sociology (Friedberg and Crozier 1980). What resources of its own can the P2A draw upon (budget, specific service) in undertaking its cross-cutting endeavours and inducing the other services and elected officials to comply with its recommendations? The human and financial resources made available for this public policy will be an important indicator of its effectiveness.

All of which leads us to ask ourselves: what will the governance of the food issue look like at the metropolitan level? And what then of the role of civil society actors? Whereas in some cities, like Lyon (Brand 2015), the social movement has driven local public action, such has not been the case in Montpellier. For now there is no strong link between food movements and P2A implementation. Montpellier has very many associations but so far their activities have garnered little support from Montpellier Méditerranée Métropole in terms of a comprehensive food policy. The policy was indeed developed without any discussion with them. Cooperation does exist between associations and public institutions in some limited spheres where it has become institutionalized: such is the case, for instance, of the cooperation between the Department of Hérault and food aid associations, or the arrangements governing collective gardens established by the City of Montpellier. While many citizens' initiatives do exist, a new stage of the P2A calls for greater openness to civil society.

Conclusion

In the case of Montpellier, if food policies are now on the agenda, that is due to the combined effect of administrative and political timetables, which have opened up a window of opportunity as defined by Kingdon (1984). However, that dynamic is both made possible and constrained by pre-existing fragmentary public actions on related issues. The second process is the framing of the food issue, which depends on the development of shared knowledge of the field, namely the urban food system. As in other cities worldwide, the approach we can observe in Montpellier involves the mobilization of scientific and technical resources, which produces its own framing effects: there are areas of interest in which local actors are overinvested, while other matters that had been neglected are brought to light by cooperation with the researchers. The third process relates to the many sociopolitical reconfigurations involved in food governance. New elected officials are the ones developing that governance, but they must deal with the political system in place. By bringing the food issue into the sphere of local political action, they help to make a subject that was formerly marginal a model of openness to political innovation. A cross-cutting concern by definition, food supply is an area where cooperation is unavoidable for administrative and technical services whose jurisdictions were initially assigned to them in terms of the key mandates of a French metropolis. It also leads to reconfigured relationships between city centres and their outskirts and between urban, periurban and rural areas. The fourth process is the mobilization of civil society actors, which the scientific literature holds to be a prime vector in the construction of sustainable food governance. Their participation, however, has barely begun in Montpellier, but also more generally throughout France, where the tradition of State and local governments' taking responsibility for public problems still holds sway.

References

Béal V, Gauthier M, Pinson G (2011) Le développement durable changera-t-il la ville? Presses de l'Université de Saint-Étienne, 457 p

Becker H (1963) Outsiders. The Free Press of Glencoe, 249 p

Bonnefoy S, Brand C (2014) Régulation politique et territorialisation du fait alimentaire: de l'agriculture à l'agri-alimentaire. Géocarrefour 89(1–2):95–103

Bonnet V (2016) Aménager les territoires ruraux et périurbains. Report submitted to Sylvia Pinel, Minister of Housing and Territorial Equality, 129 p

Brand C (2015) Alimentation et métropolisation: problématique vitale oubliée. Doctoral thesis in geography, université Grenoble Alpes:656 p

Cobb R, Elder C (1972) Participation in American politics: the dynamics of agenda-building. Allyn and Bacon, Boston, 182 p

Darré JP (2006) La recherche co-active de solutions entre agents de développement et agriculteurs. Éditions Gret, Cnearc, Gerdal, 52 p

Derkzen P, Morgan K (2012) Food and the City: the challenge of urban food governance. In: Viljoen A, Wiskerke JS (eds) Sustainable food planning: evolving theory and practice. Wageningen Academic Publishers, Wageningen, 598 p, pp 61–66

Friedberg E, Crozier M (1980) Actors and systems: the politics of collective action. University of Chicago Press, Chicago

Friedmann H (2007) Scaling up: bringing public institutions and food service corporations into the project for a local, sustainable food system in Ontario. Agric Hum Values 24(3):389–398

Giulani M (1999) Sul concepto di imprenditore di policy. Rivista Italiana di Scienza Politica 28(2):375–378

Jarrige F, Thinon P, Nougaredes B (2006) La prise en compte de l'agriculture dans les nouveaux projets de territoires urbains: Exemple d'une recherche en partenariat avec la communauté d'agglomération de Montpellier. Revue d'Économie Régionale et Urbaine 3:393–414

Kingdon J (1984) Agendas, alternatives and public policies. Little Brown, Boston, 240 p

Lefèvre C (1998) Gouvernements métropolitains et gouvernance dans les pays occidentaux. Politiques et management public 16(1):35–59

Le Velly R, Paturel D (2013) Des circuits courts pour l'aide alimentaire? Hybridation de régulations dans un marché expérimental en Languedoc-Roussillon. Revue d'Études en Agriculture et Environnement 94(4):443–465

Marks G (1996) An actor-Centred approach to multi-level governance. Reg Fed Stud 6(2):20–40

Michel L, Fouilleux E, Bordier L (2014) Sustainable food governance in urban areas. The Case of Montpellier (France), Montreal, 19–24 July. http://paperroom.ipsa.org/papers/paper_35693.pdf

Perrin C, Jarrige F, Soulard C-T (2013) L'espace et le temps des liens ville-agriculture: une présentation systémique du cas de Montpellier et sa région. Cahiers Agricultures 22(6):552–558

Perrin C, Soulard CT (2014) Vers une gouvernance alimentaire locale reliant ville et agriculture. Le cas de Perpignan. Géocarrefour 89(1–2):125–134

Pinson G (2006) Projets de ville et gouvernance urbaine. Revue française de science politique 56(4):619–651

Pinson G (2009) Gouverner la ville par projet, Urbanisme et gouvernance des villes européennes. Presses de Science Po, Paris, 418 p

Pothukuchi K, Kaufman J (2000) The food system: a stranger to urban planning. J Am Plan Assoc 66(2):113–124

Scheromm P, Soulard CT (2018) The landscapes of professional farms in mid-sized cities, France. Geogr Res 56(2):154–166

Soulard CT, Vonthron S, Bricas N, Debru J, Jarrige F, Le Velly R, Michel L, Muepu A, Sandiani S, Sebbane M (2015) Construire une politique agricole et alimentaire pour la métropole de Montpellier. Preparatory study. INRA/Montpellier Méditerranée Métropole report

Stierand P (2012) Food policy councils: recovering the local level in food policy. In: Viljoen A, Wiskerke JS (eds) Sustainable food planning: evolving theory and practice. Wageningen, Wageningen Academic Publishers. , 598 p: 67–78

Stoker G (1991) The politics of local governments. Macmillan, London, 303 p

Thinon P, Jarrige F, Nougarèdes B, Pariset G (2003) Analyse des espaces agricoles et naturels de la communauté d'agglomération de Montpellier. Unités paysagères, systèmes de productions agricoles, valeurs économiques et pratiques sociales. Volet agricole du diagnostic de Scot de la CAM, Montpellier, 50 p

Vanier M (2003) Le périurbain à l'heure du crapaud buffle : tiers espace de la nature, nature du tiers espace. / The peri-urban area : Nature's third space ?. Revue de géographie alpine 91(4):79–89

Conclusion

Nicolas Bricas

Historically, agricultural and food issues were initially inseparable but over the last century they have been dealt with as separate entities to an increasing extent. Food issues are often incorporated in agricultural policies in countries where the economy has long been dominated by the primary sector. Most populations hampered by food insecurity in these countries are rural and agriculture-oriented, and this slice of the population is proportionately in the majority. Food insecurity there is conventionally considered to be due to an insufficient food supply, so increasing agricultural production is the main strategy adopted to fight hunger.

In countries with a longer history of industrialization, food was always a major policy domain for cities until the issue was taken up by the States. The latter intervened to ensure food security (stocks, price regulation) and food safety (regulation, control), while providing the most vulnerable people with access to at least a minimum amount of quality food (food aid). States then partially withdrew their involvement when the liberalization trend took off in the late twentieth century and they handed over the task of fulfilling urban food supplies to agrifood logistics, processing and distribution companies. Meanwhile, States maintained control over food health safety. Food security in these countries was relatively well maintained from a quantitative standpoint, but State agricultural policies continued to support agriculture, which in turn became more of an economic sector whose competitiveness had to be bolstered in an aggressive international environment.

The pattern of implementation of agricultural and food policies at the urban level has actually differed from that of other urban policies. These two 'worlds' have more or less ignored each other, leading to a clearcut urban-rural divide. Urban policymaking on food issues—which has been taking place in both industrialized and so-called 'developing' countries in recent years—seeks to bridge this divide.

N. Bricas (✉)
UMR MOÏSA, Cirad, MONTPELLIER CEDEX 5, France
e-mail: nicolas.bricas@cirad.fr

© The Author(s) 2019
C. Brand et al. (eds.), *Designing Urban Food Policies*, Urban Agriculture,
https://doi.org/10.1007/978-3-030-13958-2_7

In this respect, it is crucial to take the territorial scale into account when drawing up urban food policies. Such policies may tend to limit city interventions solely to the immediate territory (almost entirely urbanized), with the risk of overlooking relationships with outlying agricultural areas that meet cities' food supply needs and with which they have *de facto* links. These links are generally not identified—cities are often unable to draw up maps of the agricultural areas that supply them. Conversely, urban conurbations, which encompass communities located in the vicinity of the city and where agriculture is a major territorial component, often integrate both food and agricultural issues. If from a hypothetical standpoint we consider that the urban world is made up mainly of consumers and the rural world of farmers, we still cannot assume that urban dwellers will essentially be concerned by food issues and rural dwellers by agricultural issues. The contrary pattern is revealed when looking at conurbations—which incorporate both worlds—where rural people are also concerned by food issues (e.g. food deserts and malnutrition), while agricultural issues (urban crop and livestock farming) are also on the minds of urban dwellers. This is the case in metropolitan Montpellier (France) which has implemented an 'agroecology and food policy', so these two key elements are dovetailed. Urban policymaking regarding food issues will be more relevant if territories that combine both urban and rural geographical spaces are taken into consideration. Moreover, the promotion of the city region food systems (CRFS) concept by many actors reflects an intention to bridge these rural-urban and agriculture-food divides.

Many urban food policies are currently aimed at relinking the two worlds by relocalising food production to feed cities. Some cities even intend to achieve food self-sufficiency in their territories. Historically, it is nevertheless known that large cities were built using supplies procured from remote lands. With a few exceptions, due to the size of cities, it would be wishful thinking to strive for self-sufficiency based on sourcing food supplies solely from lands within the immediate vicinity of these urban areas. Although food self-sufficiency with regard to vegetable, egg or poultry supplies could be imagined, this would be hard to achieve for other foods given the size of cities and the quantities involved—cereals, roots, tubers, legumes, meat, fish, oil and sugar, for instance, could not be solely produced in the urban hinterland. The urban food policy challenge is thus to consider food systems on a broader scale than just urban and periurban territories:

- from a geographical viewpoint, taking into account agricultural production areas that feed the city, even when they are remote, as well as processing, logistics and urban waste recycling areas;
- from a political viewpoint, because the food sector in cities is currently largely shaped by national policies, regional (CAP in Europe, NAFTA in North and Central America, ECOWAS in West Africa, etc.) and international (e.g. WTO) agreements, and by major processing, distribution and catering companies.

Urban food policy councils are usually composed of representatives of local institutions, but the question arises as to how cities could interact with non-local institutions that nevertheless have a major influence on food supplies for their communities. This is one of the issues tackled by the Milan Urban Food Policy Pact initiative, which includes over a hundred cities from around the world that are building more sustainable food system policies. The aim—through a federated city movement—is to interact with national, regional or global bodies whose remit extends well beyond the territory of individual cities.

The complex, interwoven, multifactorial and multidimensional aspects of urban policymaking on agricultural and food issues are clearcut. In this book, we opted to not present a holistic and integrative conceptual framework of all aspects of this policymaking (of varied and sometimes contradictory interest). Instead we propose a framework that reconciles the different policymaking pathways by focusing on: solving the food problems at hand; marshalling levers that local urban governments have available to tackle the food issue; or (but to a lesser extent) setting food policymaking as a primary objective. These pathways are obviously not linear, finished or definitive, instead they should be taken as an aggregate of linked combinations, as clearly highlighted by Caroline Brand in her doctoral thesis. Awareness of these formulated policymaking pathways could quickly convince urban actors that food management is essential, while helping them understand that they have actually been dealing with food in separate silos 'without knowing it'. Their mobilization on this cross-cutting issue will then be stronger than if they were to create a new integrative silo, which could stifle some of their prerogatives. The debate that is presently under way—to which this book contributes—has already pinpointed several issues that research teams specialized on this topic intend to deal with in the future.

The first issue concerns a key urban food system actor that surprisingly is not yet at the forefront in the debate—the urban population, i.e. the consumers or 'eaters'. Very little is known and published about their practices and representations, particularly to help gain insight into their impacts on food system sustainability. The literature is geared more towards supply chains and food supply organization than towards practices in the domestic sphere, i.e. households. How do urban food policies affect them? New approaches focused on the factors that determine food styles recognize the role of the physical and economic environment in behavioural changes. This innovative field of research explores the effects of the foodscape and urban food environment—which in turn is shaped by cities—on food behaviours and representations.

The second issue concerns the political role of cities in food system management at broader scales and, conversely, concerns the impacts of these national, regional or global food systems on urban policy—are they synergistic or antagonistic? What degree of latitude is there for urban policymaking on food issues?

Finally, the third set of issues concerns innovations that make effective use of urban resources. Cities give rise to specific problems regarding food system sustainability. The resources they channel could however be promoted to help solve these problems. The growing number of urban innovations concerning ways to produce or gain access to food—through trade or other alternative (e.g. collaborative) means—or even ways to cook and eat, reflects an inclination towards the gradual invention of new food systems. The challenge here is to mobilize research, no longer to conduct laboratory analyses on responses to major sustainability challenges (also defined in the laboratory), but rather to co-build and support these innovations on the basis of what motivates the actors that manage them, while assessing their effects on sustainability. Research can therefore help promote the capacity of cities to overcome food system sustainability issues and ensure that they will no longer hamper this process.